Handbook of Modern Experiments for High School Biology

Previous Books by the Author:

Question and Answer Course in General Biology
Simple Science Experiments
Modern Science Dictionary

HANDBOOK OF MODERN EXPERIMENTS FOR HIGH SCHOOL BIOLOGY

Adelaide Hechtlinger

PARKER PUBLISHING COMPANY, INC.
West Nyack, N.Y.

LIBRARY OF CONGRESS
CATALOG CARD NUMBER: 73-129545

PRINTED IN THE UNITED STATES OF AMERICA

ISBN-0-13-380451-8

B & P

How To Use This Biology Laboratory Guide

Each succeeding generation adds to our fund of scientific knowledge so rapidly that it is difficult for most teachers to keep up with all new developments or even the most important. Thus, as its title implies, it is a prime purpose of this book to give the high school biology teacher a selection of the latest materials and laboratory exercises feasible for use at the high school level. Since the material now being taught to the student is so much more sophisticated than that taught even ten years ago, it is clear that the investigations a student conducts in the high school laboratory today should also be more sophisticated than those required in the past.

One of the aims in studying biology is to understand what biology really is. The student should develop an understanding of and an interest in living organisms through certain basic concepts. This book provides the teacher with dissections and certain other exercises in the school laboratory which will enable the student to develop this understanding and interest in living organisms.

At the same time the student should develop an understanding of the scientific process and its methods of inquiry. Only by experience in the laboratory can a student see what science really is. He must do the kind of things that scientists do in the laboratory. No matter how much a student has read about the facts of science, it still remains for the laboratory to acquaint him with these experiences. Exercises such as chromatography are included to help the student develop this understanding of the scientific process.

Each exercise is so designed for the teacher that it can be set up and performed with a minimum of effort and materials. Since in most cases the student will be doing laboratory work for the first time, every exercise is presented with an introduction giving necessary background information about the work to be done. This may be developed by the teacher in class discussion prior to the laboratory, or it may be adapted to student laboratory work sheets. Each exercise also includes a list of the necessary materials to be used in conducting the experiment. Finally, it provides a detailed step-by-step procedure for performing the experiment. When necessary, questions are provided for the teacher to direct to students to help them interpret what they are doing and thus give more meaning to the topics being studied. Most of the exercises included can be done in a single laboratory session.

These exercises can be used in conjunction with any textbook. The experiments are in blocks, each group for a particular topic studied. The topic sequence is the one used by the New York State Board of Education for its suggested study of biology in the high school. At the beginning of each unit the basic understandings of that unit are given in outline form.

There may be too many experiments to do in a single year but the teacher should be able to make a good choice for his own particular needs. Some of these exercises should really be done if possible since they are the ones not usually found in the high school experiment books. They are different and the student will really enjoy doing them. A number of the exercises will not be found in other books since the author helped to work them out just prior to publication of this book.

The study of biology is undergoing great changes. Biology is no longer a topic to be studied as a separate discipline. It is now fundamentally tied in with physics and, especially, chemistry. Included in the biochemistry unit, the teacher will find many exercises very seldom performed by the high school student. These exercises will make stimulating and informative experiences for students, and they teach many skills. Too often in class the student is given complicated formulae to memorize for amino acids, sugars, proteins, and the like, without his knowing what some of these substances really are. Use of a number of these biochemistry laboratories will give a student a real understanding of their composition and function in life.

Through his experience in the laboratory, the student may be led to an understanding of biology's important role in our everyday life. He may even acquire a deep and lasting interest in the study of life. We need more biologists. Perhaps some of the following experiments will help the student in choosing his career.

ADELAIDE HECHTLINGER

SPECIAL NOTE TO THE READER

Materials must be prepared in advance for many of the experiments contained in this book. Such experiments are indicated by the asterisk after the heading.

Acknowledgments

I am grateful for the permission granted by the New York State Education Department to reprint sections from the New York State Education Department Biology Syllabus.

I would like to thank my colleague and very good friend, Nettie Herman, for her invaluable help in the preparation of this book.

CONTENTS

UNIT I

THE STUDY OF LIFE

I. THE PERSPECTIVES

A. *The Human Individual*: Each individual is one of a vast number of living things, yet has a uniqueness in the biological world.

1. *In the Universe*: Living things reside on a planet which may or may not be unique in being able to support life.

2. *Among Other Things*: Man may be considered as the highest form of life at the present time.

B. *Relationship of Biology to Other Disciplines of Science*: The study of biology requires some knowledge of physics, chemistry, earth science and mathematics.

II. THE CONCEPT OF LIFE

A. *Definition of Life*: No one, all-inclusive, definition of life has proved satisfactory. Extensive examination of many living things has, however, produced some characteristics of life which are generally agreed upon.

B. *Activities of Life*: The characteristic activities associated with living organisms are known as physiological activities or functions and include: nutrition, transport, respiration, excretion, synthesis, regulation, growth, and reproduction.

The term metabolism is used to describe the chemical activities that occur in an organism.

1. *Nutrition*: Nutrition involves those activities of an organism by which it procures raw materials from its environment and processes them so as to make them available for physiological activities.

2. *Transport*: Transport involves that intake (absorption) and distribution (circulation) of essential compounds through an organism.

3. *Respiration*: Respiration involves the conversion of the potential energy of organic compounds to a form that is suitable for the various energy-requiring activities of an organism.

4. *Excretion*: Excretion involves the removal of harmful (or potentially harmful) substances which form as waste products of metabolism in an organism.

5. *Synthesis*: Synthesis involves those chemical activities by which complex molecules are created from simple compounds.

6. *Regulation*: Regulation involves the control of the various physiological activities of an organism in such a way as to maintain stability through coordinated activity in a constantly changing environment.

7. *Growth*: Growth is the increase in size of an organism that results from the synthesis and organization of organic compounds into new materials and structures.

8. *Reproduction*: Reproduction is the uniqueness characteristic of living organisms involving the production of new individuals essentially similar to those producing them.

An organism is a living system that carries on the functions described above. It may in summary be said that organisms: (1) are complex and highly organized (even the simplest ones), (2) require and use energy, (3) are "chemical factories," (4) receive and respond to stimuli, (5) have relatively stable regulatory systems which cause certain portions of the environment to pass through them in an organized manner (principle of homeostasis), and (6) are able to reproduce their own kind.

There is some discussion among biologists as to whether viruses may be considered alive. Viruses are able to reproduce themselves if they are in living cells, but they do not perform most of the other processes that are generally considered typical of life.

C. *Life Requirements*:

1. *The Earth in the Solar System*: The earth-sun relationship, with modifying effects of atmosphere, provides special temperature and light conditions that support life on this planet.

2. *Distribution of Life (the Biosphere)*: Life on earth exists as a thin layer associated with the surface of the planet. When one considers distances in the universe, measured in light-years, the biosphere is a very thin layer of organisms interacting with their physical environment and with one another. Organisms have been found at ocean depths of about 33,000 feet (about 10,000 meters) and high in the atmosphere, but most of the earth's life exists in a narrow band between these extremes.

3. *Factors Supporting Life*: Wherever organisms exist they are interacting with one another and are limited by their physical and chemical environment.

a. *Sun*: Light from the sun is the primary source of life. The infra-red, and visible light portions of the solar spectrum furnish energy needed for photosynthesis and the temperature conditions within which life can exist.

b. *Chlorophyll*: The chlorophyll molecule has the ability to receive and convert light energy to chemical energy in green plant material. The chemical energy stored in green plant material is the source of most energy for the living world. Chemosynthesis is an exception.

c. *Water*: Approximately 70 percent of the earth's surface is covered with water. The water bodies of the earth serve as: (1) Heat "reservoirs" that regulate earth temperatures, (2) Sources of water vapor in the atmosphere, (3) Habitats and water sources for organisms.

All organisms take in, contain, and utilize water. The water taken in eventually leaves the organism and returns to the environment—an example of the environment passing through the organism.

d. *Atmosphere*: The atmosphere is an envelope of gases about the earth that acts as: (1) A reservoir for oxygen, nitrogen and carbon dioxide, (2) A "greenhouse" for organisms.

Ozone absorbs ultraviolet radiation which would cause damage to land organisms. Water vapor absorbs infrared radiations, a factor in maintaining proper temperature ranges for organisms.

e. *Earth's Crust*: The earth's crust provides a mineral supply for living organisms.

III. THE UNITS OF LIFE

A. *Cell Theory*: During the 18th and 19th centuries biologists extensively examined a wide variety of living things with the aid of the light microscope. Data collected from these investigations served as the basis for the statement of the cell theory.

1. *Currently Accepted Concepts*:

a. Cells are the unit of structure of living things.

b. Cells are the unit of function of living things.

c. Cells arise from living things.

2. *Exceptions to Concepts*: Two exceptions to these concepts which do not, however, destroy the usefulness of the theory, are:

a. The first cells could not have arisen from a previously existing cell.

b. There may be other units of structure and function. (1) Self-duplicating structures such as the mitochondria and chloroplasts, (2) Modified organizational patterns such as the multinucleate organization of slime molds, bread molds, and certain types of muscle cells, and (3) Viruses.

B. *Cell Study*: The cell is the basic unit of life. The vast majority of living things are either single cells or are comprised of many cells. Organisms function by means of their cells.

1. *Investigational Techniques*: In the more than 100 years since the statement of the cell theory, biologists have continued to study cells and, as new techniques and instrumentation have been developed, increased understanding has resulted.

The development of staining technology in the latter half of the 19th century made possible the study of structures not previously observed.

New and improved instruments have greatly aided understanding of cell anatomy and physiology. The compound microscope, electron microscope, phase-contrast microscope, interference microscope, ultracentrifuge, and microdissection instruments are some of the tools that have aided biologists in gaining a more complete picture of the structures and functions of cells.

2. *Structure and Function of Cells*: In cells, various specialized functions occur in substructures known as organelles. Some major organelles and their related functions are:

a. Cell Wall: Protects; supports plant cells and maintains constant shape; limits passage of materials into and out of plant cells.

b. Cell and Nuclear Membranes: Controls transport of materials into and out of the cells and nucleus; isolates cellular and nuclear materials.

c. Cytoplasm: Provides a watery environment in which most life activities take place.

d. Endoplasmic Reticulum: Provides channels, through which transport of materials occurs in the cytoplasm, and interfaces, where interactions may occur.

e. Ribosomes: Act as sites of protein synthesis.

f. Mitochondria: Act as sites of cellular respiration.

g. Vacuoles: Act as reservoirs for water and dissolved materials.

h. Chloroplasts: Contain energy-accumulating pigments in plant cells; act as sites for synthesis of organic compounds from inorganic materials.

i. Centrosome: Contain centrioles which are functional during the division of animal cells.

j. Nucleus: Acts as information center for cell division and other cell activities.

k. Chromosomes: Contain the hereditary material of a cell (DNA); are the agents of distribution of hereditary information.

l. Nucleolus: May have a role in protein synthesis.

There is a basic similarity of structure and function in plant and animal cells. Some unicellular organisms have characteristics of both plant and animal cells. Euglena, having characteristics of both plant and animal cells, may represent a link between plants and animals. Some differences are found between plant and animal cells. Structures unique to each type of cell are criteria used to distinguish between plant and animal cells.

3. *Size Relationships in Cell Study*: Appropriate units like the micron

(μ) are used in cell measurement. The use of larger units in discussing sizes of cells and their organelles is cumbersome (e.g., the diameter of a red blood cell is more conveniently expressed as 7 μ than as 0.0000002 feet). One micron is equal to .001 mm.

a. *Organelle size*: The organelles are small as compared to the size of the cell.

b. *Relationship of Size to Function*: The size of cells and organelles is related to the task they perform and is not haphazard. Cells vary in size from bacteria (0.2 μ) to the ostrich egg (75,000 μ). The smallest cell now known is PPLO (pleuro-pneumonia-like organism) which is 0.1 μ.

CARE AND USE OF THE COMPOUND MICROSCOPE

AIM: To learn how to use the compound microscope properly and develop skill in preparing materials for use with the microscope.

The microscope is a basic scientific instrument especially designed to use in the study of objects that are too small to be seen or studied by the eye alone. The microscope brings a whole new world to the student.

The student usually uses a compound light microscope that magnifies from 100 to 440 times. There are other types of microscopes that the student might learn to use before the course is over but the compound microscope is most important.

There is the binocular dissecting microscope, often called the stereoscopic dissecting microscope, which magnifies from 4 to 40 times. There is also the oil-immersion lens and the phase-contrast microscope.

Of course there is the electron microscope which magnifies more than 100,000 times but this microscope is very expensive and used only by highly trained technicians in very large laboratories and universities for research.

MATERIALS NEEDED: Microscope, microscope slide, cover slip, cheesecloth, lens paper, piece of newsprint, medicine dropper, water, pair of forceps or dissecting needles, pencil with eraser.

PROCEDURE:

(A) Preparation of Microscope for Use:

1. When taking the microscope from its storage place, carry it with both hands. Hold the arm of the microscope with one hand and place the other hand under the base. Carry it in an upright position and do not tilt it.

2. Set the instrument down gently with the arm toward you and the base two or more inches from the edge of the table. The microscope should be kept in an upright position and not tilted at the inclination joint.

3. Use the diagram so that you are familiar with the various parts of the microscope and the function of each part.

4. Use a soft cloth, such as cheesecloth, to wipe off any dust from the face of the mirror, the stage and other parts of the frame. *Do not use the cloth for the lenses.*

5. Use a piece of lens paper to wipe the top lens of the ocular and the tip lens of each objective. The usual student model compound microscope magnifies from 100 to 440 times.

6. Rotate the nosepiece so that the low power objective (the shorter one) is in line with the body tube. When it is in the proper position it should either click or snap into place.

7. Open the diaphragm as if it is an iris, or turn the disk to the largest opening so that the greatest amount of light is admitted.

8. Put your eye to the eyepiece and move the mirror around until it reflects light upward through the opening in the stage. Do not use the flat side of the mirror but use the concave side. Make certain that the mirror is not directly toward the sun or the illumination will be too bright.

9. Continue looking through the ocular of the microscope and adjust the mirror, if any, and the diaphragm so that the round field of view is evenly illuminated and there is no glare. The field is a uniform circle of light with no shadows. If you see any smudges, wipe the lenses again with a fresh piece of paper. When you have a clear field of light, the microscope is ready for use.

(B) Using the Microscope:

Materials to be studied under the microscope are mounted on a rectangular glass slide and are covered with a thin round or square piece of glass or plastic covered with a cover glass or cover slip.

1. Slides and cover glasses should be held by their edges to avoid getting smudges or fingerprints on the surface. Always wash the slide and cover glass and wipe both sides carefully. Make certain that the cover glass is clean on both sides.

2. Cut a single letter "e" from the smallest newsprint you can find and lay it right side up in the center of the slide.

3. With a medicine dropper put one drop of water on the piece of paper.

4. Wait a few minutes to allow the paper to soak up the water and then hold the cover glass at a 45° angle to the drop of water and lower it gently over the materials. Use a dissecting needle or a pair of forceps to lower the glass gently over the materials so that there are no air bubbles trapped under the cover. If any bubbles are found, tap the glass gently with a pencil eraser and the bubbles will move to the edge of the cover glass where they can then escape. You now have a wet mount ready for microscopic examination.

5. Place the mounted slide on the microscope stage so that the paper is centered above the opening in the stage.

6. Now look at the microscope from one side and, watching the bottom lens, turn the low-power objective down until it almost touches the cover glass or until it will automatically stop. NEVER LOWER THE OBJECTIVE WHILE LOOKING THROUGH THE EYEPIECE—YOU MIGHT ACCIDENTALLY LOWER IT TOO MUCH AND BREAK THE SLIDE AND COVER GLASS.

7. Keep both eyes open while looking through the ocular. It might be difficult to do this at first but the use of only one with the other closed might cause eyestrain. Right-handed students should use their left eye and left-handed students, their right eye. They will find it is much simpler when drawing what they see through the microscope.

8. Put your eye to the ocular and, as you watch the field, use the coarse adjustment to raise the body tube until you see the printed letter. For the ordinary student compound microscope in proper focus, a 10× low-power objective should be about 5/8″ (16 mm) above the material. The microscope should only be focused by raising the body tube.

9. Use the fine adjustment to bring the letter into sharper focus. Examine the letter by turning the fine adjustment slowly back and forth, not more than half a turn. This will cause the focus to shift and you will see more details at different levels.

10. TO USE THE MICROSCOPE'S HIGH POWER: Move the letter to the exact center of the low-power field and turn the high-power objective into position. Make certain to look at the microscope from one side so that the objective does not hit the slide. Look through the ocular and use the fine adjustment only to bring the material into proper focus. NEVER USE THE COARSE ADJUSTMENT FOR HIGH POWER.

11. Make a drawing of what you saw under low power and what you saw under high power.

QUESTIONS:

1. Compare the position of the letter "e" as you saw it in the paper with its appearance in the microscopic field.

2. If you moved the slide on the stage from right to left, how is the position of the material changed in the microscopic field?

3. When viewing the material under high power, does the field of view show a larger or smaller area of the object?

4. Is the brightness of the field greater or less than with the low power?

ADDITIONAL EXAMINATIONS:

Examine blonde, brunette, gray and red hair for pigments. Make wet mounts of the hair. Also examine curly hair, straight hair, and permanently-waved hair for shape in the same manner. Have each student do a different slide so that he or she will then be able to look into other microscopes to see the material.

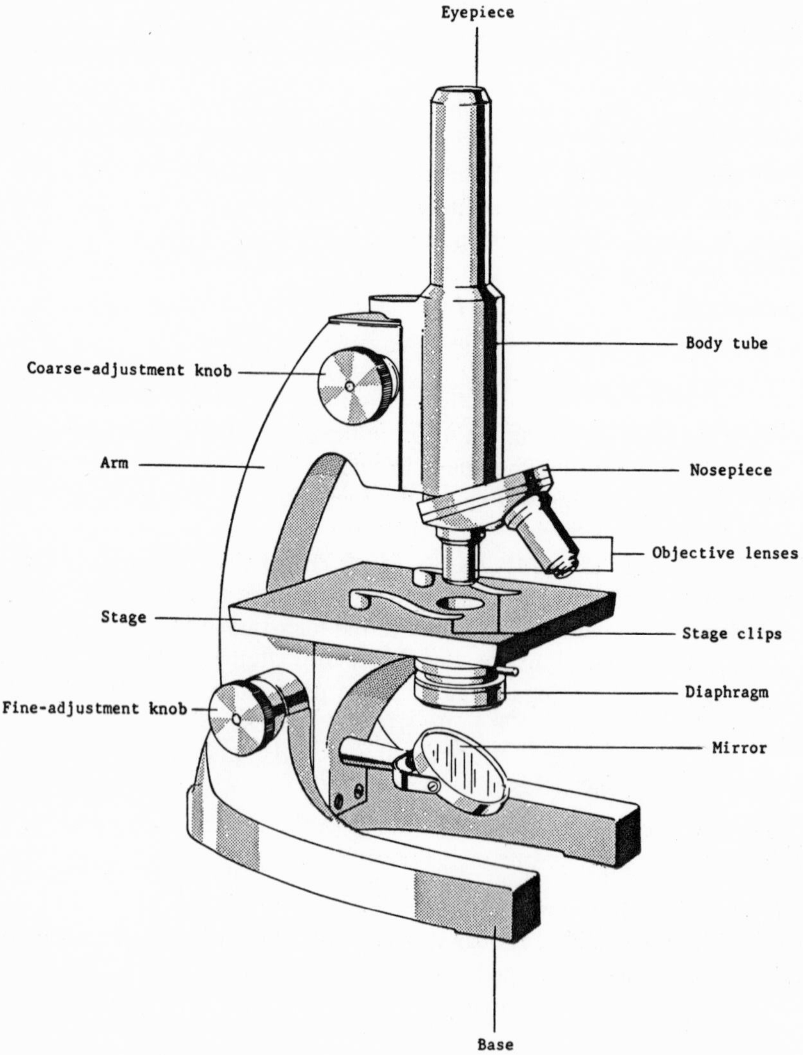

Diagram 1. A Compound Microscope.

PARTS OF THE MICROSCOPE

OCULAR OR EYEPIECE: Fits into the top of the body tube. It contains the lenses to increase magnification. The eyepiece usually has a magnification of 10× but it may be replaced with another of lower or higher magnification.

BODY TUBE: It holds the eyepiece at one end and the objective at the other end. It holds them at the proper working distance from each other.

COARSE ADJUSTMENT: The body tube can be moved quickly up or down. It should be used only when the low-power objective is in place. It moves the body tube approximately to the correct distance from the specimen.

ARM: Supports the body tube and coarse adjustments.

NOSEPIECE: Used to interchange the low- and high-power objectives.

OBJECTIVE: Contains lenses which magnify the material being examined. The lenses are of different magnifications. In the ordinary student compound microscope, the shorter one is the low-power, 10× objective and the longer is the high-power, 43× objective.

STAGE: This is the platform on which the materials are placed to be examined. The slide is placed over the hole that admits light from the mirror below.

DIAPHRAGM: Regulates the amount of light passing through the specimen by opening or closing a screen of metal plates.

STAGE CLIPS: Hold the specimen slide firmly in place.

BASE: Support that holds the weight of the microscope.

MIRROR: The mirror directs light through the material being studied. The mirror has one flat and one concave surface.

INCLINATION JOINT: Permits tilting to adjust to eye level.

FINE ADJUSTMENT: A very delicate adjustment of the level of the body tube can be made. A sharper focus can be obtained.

MEASUREMENTS WITH A MICROSCOPE

AIM: To learn how to use the metric system for making measurements of objects viewed under the microscope.

Objects viewed under the microscope are so small that you cannot use direct measurement. However, you still can measure the object seen under the microscope by estimating the diameter and area of the low- and high-power fields of your microscope. In this way you will be able to determine indirectly the approximate size of the object. There are several different methods that can be used.

Microns are the most common unit of measurement in microscope work. A micron (μ) is equal to 0.001 mm. In addition you should know some units and subunits of the metric system.

SOME UNITS AND SUBUNITS OF THE METRIC SYSTEM

Greek Fraction Prefixes	*Applied to a Meter*
$\frac{1}{\text{tenth}}$ = deci	$\frac{1}{\text{tenth}}$ meter = 1 decimeter (dm)
$\frac{1}{\text{hundredth}}$ = centi	$\frac{1}{\text{hundredth}}$ meter = 1 centimeter (cm)
$\frac{1}{\text{thousandth}}$ = milli	$\frac{1}{\text{thousandth}}$ meter = 1 millimeter (mm)
$\frac{1}{\text{millionth}}$ = micro	$\frac{1}{\text{millionth}}$ meter = 1 micron (μ, Greek letter "m")
$\frac{1}{\text{billionth}}$ = millimicro (a thousandth of a millionth)	$\frac{1}{\text{billionth}}$ meter = 1 millimicron (mμ)

MATERIALS NEEDED: A plastic ruler, microscope, object to be measured.

PROCEDURE:

1. Although you can easily calculate the amount of magnification produced by your microscope by multiplying the magnification of the ocular lens by the objective lenses, to determine the size of the object being viewed it can be compared to another microscopic object of known size in the same field. This can be done by placing a plastic millimeter ruler on the stage across the midline of the microscopic field. The object to be measured is then placed next to the ruler and the size of the object can be determined directly in millimeters or fractions of a millimeter.

2. However, there is a more convenient way to measure objects. You measure the diameter of your microscope field, and then you can estimate the size of the object in that field. For example, if your field diameter is 1200 μ and the object extends a third of the way across the field, you can then estimate that the object is approximately 400 μ in width.

3. Another method to determine the size is as follows: Use the low-power and place the graduated edge of a plastic metric ruler across the midline of the field of vision. Focus the microscope so that the markings on the ruler can be clearly seen. Record the diameter of the field in millimeters.

From this figure, you can calculate the area of the field since the area of a circle is equal to πr^2 where $\pi = 22/7$ and r is ½ the diameter. Now record the diameter in microns.

4. To calculate the diameter of the high-power field, divide the high-power magnification by the low-power magnification to get a factor that shows how much smaller the high-power field is. If the low-power objective reads 15× and the high-power reads 60×, dividing 60 by 15 will give a factor of 4. If you had calculated the diameter of the low-power to be 1.6 mm then the diameter of the high-power field would be 0.4 since 1.6 divided by 4 will be 0.4.

Thus since you know the diameter of the low-power and the high-power of your microscope field, you can estimate the size of the object you are viewing under the microscope.

5. Make a slide of a hair and estimate its diameter.
6. Make a slide of a letter from the newsprint and estimate its size.

VOLUME AND WEIGHT MEASUREMENTS IN BIOLOGY*

AIM: What is the effect of varying concentrations of salt upon the volume and weight of a piece of potato?

INTRODUCTION: In this exercise you will gain experience in the use of laboratory apparatus for measuring volume and weight. At the same time you will observe changes in living material when subjected to varying concentrations of salt.

MATERIALS NEEDED:

For Teacher: 1 liter graduated cylinder; empty quart milk container.

For Each Group: Jars A, B, C, containing distilled water, 5% NaCl, and 10% NaCl, respectively. Each jar labeled with measurements of potato core before immersion; 3 metric rulers; 10 ml. graduated cylinder; triple beam balance; weighing paper; dissecting needle paper towels; potato cores.

PROCEDURE: Each group of students is supplied with three jars, each of which contains a cylindrical piece of potato immersed in a liquid. On the previous day these pieces of potato had been cut from the potato by means of a cork borer. Measurements were made of the cores of the potato and the cores placed into the fluid in the jars. The data is on the label of each jar. You will repeat these measurements in order to determine the effect of a 24-hour immersion in the liquid. Within each group measurements of each core should be made simultaneously.

* Starred experiments throughout the book should be checked for materials at least two days in advance. Some materials must be prepared ahead of time.

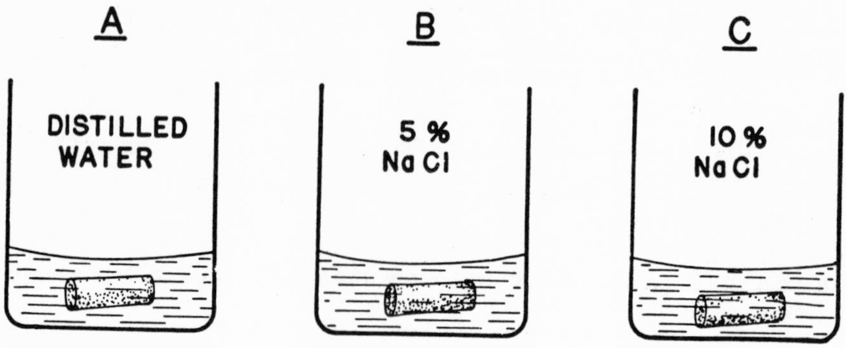

Diagram 2.

1. From the labels on each jar, copy onto the tables included in these directions the data pertaining to the cores before they were immersed.

2. Volume By Calculation:

a. Measure the length and diameter of the core to the nearest millimeter. Record your data.

b. Calculate the volume in cubic centimeters (cm^3 or cc).

3. Volume by Direct Measurement: The volume of the core may also be directly determined by its displacement of water. Measurements will be made in milliliters. To give you a concept of the *liter*, the unit of fluid volume, your teacher will pour water from a one-quart milk container into a one-liter graduated cylinder.

Pour water into the graduated 10 ml cylinder until it is about half full. Record the volume in milliliters, reading to the bottom of the meniscus and holding the meniscus at eye level.

Hold the core by a needle and insert it into the water until it is completely submerged. Record the water level again. Subtract the previous reading to determine the volume of the core in milliliters. Record the volume.

4. Weight: Place paper on the pan of the balance and *tare* the balance as demonstrated to you.

Roll the core on paper toweling and weigh the core to the nearest tenth of a gram. Record your data.

5. Exchange data so that each student in the group has information about cores A, B, and C.

OBSERVATIONS AND CALCULATIONS:

CORE A—DISTILLED WATER			
Measurement	before immersion	after immersion	difference
Length, cm			
Diameter, cm			
Volume, cm^3			
Volume, ml			
Weight, g			

CORE B—5% NaCl			
Measurement	before immersion	after immersion	difference
Length, cm			
Diameter, cm			
Volume, cm^3			
Volume, ml			
Weight, g			

CORE C—10% NaCl			
Measurement	before immersion	after immersion	difference
Length, cm			
Diameter, cm			
Volume, cm^3			
Volume, ml			
Weight, g			

DISCUSSION: Answer the following questions.

1. About how many liters are there to one quart?
2. What is the volume of a liter in cubic centimeters?
3. About how many cubic centimeters to one milliliter?
4. You have used two methods for determining the volume of the potato cores: a calculation giving volume in cm^3 and a direct measurement giving volume in ml. How do the two numbers compare with each other?

5. What changes were there in Core A (distilled water) after immersion for 24 hours?

6. How does the change in volume compare with the change in weight?

7. What changes occurred in Core B? Core C?

8. What relationship is there between the concentration of salt in the solution and the change in the potato cores?

9. Why would it be reasonable to form the following hypothesis?—There should be some salt concentration in which the potato cores would not change.

10. Describe an experiment that would test this hypothesis.

LIVING AND LIFELESS SYSTEMS

AIM: To study a lifeless system and see what properties of organisms it can imitate.

Often a lifeless system in its mannerisms might show characteristics which are almost living.

MATERIALS: Mercury, nitric acid; potassium dichromate solution; petri dish; iron nail; and a dissecting needle.

PROCEDURE:

(Caution: the materials used in this exercise have fumes that can be dangerous. Therefore, do not touch the materials more than necessary. Do not put your face close to them and do not remove the materials from the dish.)

1. Place a small quantity of mercury into the dish. The mercury should be about the size of a dime. Pour enough nitric acid over the mercury to cover it completely and then add 20 drops of the potassium dichromate solution. Place the tip of the nail into the mercury. When you do this, the mercury will look as if it has a steady beat.

2. The reaction rate and size can be regulated by adding various amounts of acid and dichromate, and stirring the mixture with a dissecting needle before inserting the needle into the mercury.

3. You will notice that the mercury has an ameboid movement. When the activity slows down, more reagent will start it up again.

4. When the exercise is over, wash the mercury by placing it under running water. Be certain that the mercury remains in the dish, and then return the mercury to its container.

CONCLUSIONS:

What characteristics have you noticed that are like the living?

ROBERT HOOKE'S CELLS

AIM: To see the same type of cell that Robert Hooke saw when he named them cells.

Three hundred years ago, Robert Hooke, an English naturalist and inventor, first saw and named plant cells. In this exercise you will repeat Hooke's experiment, somewhat improved since your microscope is not as crude as his was. However, you will still use the same material, cork.

Cork is the outer layer of the bark of an evergreen oak, which grows chiefly on poor, rocky soil in the countries along the Mediterranean. Cork is used for bottle stoppers, for fishing floats, for life preservers and for lifeboats because it is almost impervious to liquid.

MATERIAL USED: Compound microscope, single-edged razor blade, bottle cork, slide and cover glass, pipette or medicine dropper.

PROCEDURE:

1. Hold the cork firmly on a flat surface and use the razor blade to cut a thin slice from the end of the cork.
2. Also cut a thin slice from the side of a cubical piece of the cork. The slices should be very thin, almost transparent.
3. Place a section on a clean slide in a small drop of water and add another drop.
4. Cover carefully with the cover glass so that there will be no air bubbles.
5. Place slide on microscope stage and observe the specimen under low power.
6. Draw what you see.
7. Answer the following:
   a. Are all the cells the same size and shape?
   b. Do the cells have spaces between them?
   c. What structures do you see inside each cell?
   d. What part of the cell would keep out water when cork floats?
   e. Why would cork float?

PLANT CELLS

AIM: What do plant cells look like?

You have seen Hooke's cork cells. Of course, they are dead cells. Now you will look at live plant cells and compare them to the cork cells.

MATERIALS NEEDED: Compound microscope; an onion; iodine; pipette or medicine dropper; microscope slide; cover glass; paper toweling; forceps; dissecting needles; and razor blade.

PROCEDURE:

1. Cut a slice from an onion. Select one of the thick, juicy, inner layers.
2. With the tweezers, peel off a piece of the inner skin of this layer. This is the epidermis. This skin is very thin, and it should come off in an even sheet.
3. Put the piece of skin in a drop of water on a clean glass slide. Use two dissecting needles to tease the edge of the preparation until the layer of tissue is completely flat and does not overlap.
4. Lay the cover glass over the epidermis and wipe off the excess water with the toweling.
5. Examine the slide under the low power of the microscope. You should see the cells which are bricklike in shape. You should also be able to see the cell walls clearly. Look for a tiny, round, yellowish lump inside each cell. This is the nucleus of the cell. Also look for the clear protoplasm around the nucleus or in little streaks near the cell walls or across the cells. It will not be easy to see.
6. Remove the slide from the microscope and gently lift the cover glass from the preparation. With a piece of paper toweling blot the water from the slide and straighten out the preparation again. Add a drop of iodine and allow the preparation to stand a moment. Carefully replace the cover glass so that there are no air bubbles and examine the preparation under low power again. You can now see everything much more clearly because the iodine has stained the cell structures a brown color.
7. To examine in greater detail, use the high power of the microscope. Make certain the section you wish to examine under high power is directly in the center of the field. Now turn the nosepiece to the high-power objective. Be careful not to crush the slide; follow the rules given in the microscope lesson. Sharpen the focus if necessary. Draw what you see.
8. If you have done the exercise on the measurement of the microscope field, you should be able to estimate the average length of one cell to the nearest tenth of a millimeter. Since there are usually 7 cells in the diameter of the low-power field, the width of the field should be divided by 7.

ANIMAL CELLS

AIM: What do animal cells look like?

After studying the cork cells that Robert Hooke had studied and after studying plant cells, you are about to study animal cells. You should clearly be able to see that there are several basic differences between the two.

You have seen that the surfaces of an onion bulb scale are covered with a flat layer of epidermal cells. Animals also have parts of their bodies covered with an epidermis. The inner surface of your mouth and the external parts of the body have epidermal cells.

In this exercise you will scrape a few loose epidermal cells from the inside of your cheek to be used for examination and then use skin cells from a frog.

MATERIALS NEEDED: Compound microscope, toothpick, iodine solution, methylene blue, Ringer's solution, 2 slides and 2 cover glasses, medicine droppers or pipettes.

PROCEDURE:

(A) Human Epithelial Cells

1. Place a small drop of water on a clean slide.
2. Gently scrape the inside of your cheek with the broad end of a toothpick until there is a little scraping on the toothpick.
3. Deposit a little of the scraping in the drop of water by rolling the toothpick around in it.
4. Use the toothpick to pick apart the mass of cells and stir until it is smooth.
5. Add a drop of the iodine solution to the material on the slide.
6. Cover gently with the cover glass so that no air bubbles are to be found on the slide.
7. Observe under low power of the microscope.
8. When you have found some cells answer the following:
   a. Is the cell any different externally from that of the plant?
   b. Does the animal cell have a cell wall?
   c. How do the cells fit together?
9. Switch to high power and look for the nucleus after you have one cell completely in view. What is the ratio of the diameter of the nucleus to the diameter of the cell?
10. Draw what you have seen under low power and high power and label.

(B) Frog Epithelial Cells

Since frogs continually shed their epithelium in the water, prepare a slide from the cloudy water in which a frog has been sitting.

1. Take a drop of the water and place it on the slide. You should have tiny shreds of the skin in the water.
2. Gently blot off the excess water and add a drop of Ringer's solution and then a drop of methylene blue.
3. Cover carefully with the cover glass. Make certain that there are no air bubbles on the slide.

4. Examine slide under low power and then high power of the microscope.

5. Answer the following:

a. Are the cells of the frog larger or smaller than those of the human epithelium?

b. Are the cells similar in shape?

THE LIVING CELL

AIM: How to make a silicone cell on a microscope slide that will keep living material in a condition so that it could be used for a length of time.

Several large chemical companies have been experimenting with various types of silicones that could be used as a seal for living materials. It has been found that a silicone gum is very useful as an aid for keeping such materials alive for a length of time within a small enclosed area.

It also has been found that a cell could be fashioned from silicone gum which would allow gases such as oxygen and carbon dioxide to pass through the gum and at the same time, retain the organism in its culture material in a viable state. This is most useful when working with organisms that cannot be handled on an ordinary microscope for study under the microscope because of their size. Organisms such as hydra, planaria, daphnia, nematodes and other organisms of similar type can be studied very easily. The organism can carry on all its life functions while being observed under the microscope and, if need be, the study can be continued several days later.

MATERIALS NEEDED: Silicone gum, microscope slide, cover glass, cork borer, organism to be studied.

PROCEDURE:

1. A small piece of silicone gum, the size of a pea is taken and rolled into a ball.

2. It is then placed in the center of a clean microscope slide. The gum will flow very slowly and form a disk within a few hours.

3. A #5 cork borer is used to cut a depression in the center of the disk, and the center silicone section is removed, leaving a doughnut-shaped ring on the slide. The size of the cell depends upon the size of the cork borer used. Make certain that the size of the ring is larger than the material to be studied.

4. The material to be examined is placed in the ring and then a cover glass is placed over the ring and material. Press slightly down on the cover glass. This creates a very tight seal and the slide can be inverted or put into any position without fear of the liquid leaving the slide.

5. Place the slide and material under the low power of a compound microscope, or under a dissecting scope, or use a hand lens to study the material.

6. If the material is to be used another day, place the slide with the organism in a shallow tray of water until needed. When re-used, gently wipe the water, etc. off the slide and cover glass before studying specimen again.

SLIME MOLD

AIM: To study some of the characteristics of living matter in the slime mold *Physarum polycephalum.*

The slime mold is one of the simplest forms of plants. It has masses of protoplasm, many nuclei, and no internal divisions so that it is in reality one large cell with many nuclei. The plants are found in cool, damp, shaded areas of woods where they grow on decaying vegetation.

At some times, the slime mold may seem to be animal because it moves about with an ameboid movement and looks like a giant ameba. This particular stage is called *plasmodium.* At other times the plasmodium produces fruiting bodies which form thick-walled spores. These spores develop into the organism when they alight on the proper growing medium.

MATERIALS NEEDED: A culture of slime mold, a petri dish containing nonnutrient agar, acetic acid, dissecting needle, Bunsen burner or other source of heat, glass rod, oatmeal, razor blade, slide and cover glass, compound microscope, dissecting microscope, detergent solution.

PROCEDURE:

1. Place some dried plasmodium on a small square of filter paper and then place the paper in the center of a petri dish that contains a thin layer of nonnutrient agar medium in it. Put a few drops of water on the paper. Cover the dish and keep in a well-lighted cool place for several days.

2. Within a few days jellylike growths will be seen. Under a microscope you will see that there is a constant streaming of cytoplasm throughout the mass. There are no cell walls and many nuclei.

3. If you apply a little heat to one side of the dish, and then look under the microscope, you will find that the streaming is faster if heat is applied.

4. Find a large branch of the plasmodium and puncture it with the needle. A hole will appear which soon will be repaired so that no hole can be found.

5. Use the razor blade to cut the plasmodium in several pieces. In some cases cut through the agar and in other cases, just cut the plasmodium. Watch what happens. If the agar has not been cut, the small piece of the

plasmodium will join back with the main body. If the agar has been cut, the pieces will remain independent. Thus the main body of the mold is an individual and the cut pieces are an individual when they join; it is again only one individual.

6. Place a small amount of oatmeal on one side of the plasmodium. Not too much will take place immediately since the slime mold does not react to the oatmeal. However, within 30 minutes, the slime mold will be over the oatmeal since it eats the bacteria growing on the oatmeal.

7. If you dip a glass rod in acetic acid and touch the plasmodium, the part touched will die. This can be proven by marking an X on the outside of the dish beneath the point touched with the rod and acetic acid and observe.

8. Place the dish under a dissecting microscope and look at the fruiting bodies. Put a fruiting body on a slide, add a drop of detergent solution and cover with a cover glass. Use the high power of the compound microscope to see these rough-coated spheres.

9. The plasmodium can be re-used at a later date if the filter paper containing the mold is left to dry. The dried material is good for several years. To regrow, place the dried material in a petri dish lined with several layers of filter paper, saturated with tap water and next to a pinch of dry oatmeal. Cover the dish with a glass plate and place it in an incubator at 25°C. As the mold grows add more oatmeal to the dish and keep it moist.

MOVING PROTOPLASM

AIM: To observe movement of the protoplasm in plant cells.

In addition to the plant cell being different from the animal cell in that it has a cell wall, there are other things that distinguish a plant cell. If you watch carefully you will see that the protoplasm in a plant cell has movement and that there are green bodies in certain plant cells. The green bodies are called the chloroplasts and are necessary for the plant to carry on photosynthesis. The movement of the protoplasm is due to its metabolic activity.

MATERIALS NEEDED: Microscope; silicone gum; microscope slides; cover glasses; Vallisneria (eel grass) or spinach lead, Elodea, vitamin B_1 solution; forceps; dissecting needles.

PROCEDURE:

(A) Spinach Leaf or Vallisneria:

1. Prepare a living cell with silicone gum as previously described.

2. Prepare a vitamin B_1 solution by dissolving one vitamin B complex vitamin in a half liter of water.

3. Peel a thin slide of epidermal tissue from a fresh leaf of spinach or Vallisneria.

4. Place several drops of the vitamin solution onto the living cell and add the epidermal tissue.

5. Observe under the low power of the microscope for at least five minutes and then you should make the following observations:

a. What are the green bodies?

b. Check and see what other cell structures you can find. You should be able to see the cell wall, vacuole, nucleus, and cytoplasm.

c. You should see that the chloroplasts are being carried along by streaming protoplasm. Compare their directions and speeds.

(B) Elodea:

1. Remove a fresh green leaf from an elodea plant.

2. Mount the whole leaf in a silicone gum living cell along with some of the water that the plant was growing in.

3. Hold it in your hands a few minutes to warm it.

4. Examine the cell under the microscope.

5. You should be able to see the same type of streaming as you saw in part (A). Why?

6. Draw what you saw in both parts. Label completely.

PAPER CHROMATOGRAPHY

AIM: How to make a very simple chromatogram.

Chromatography is a rapid method of separating and identifying the components of a mixture without destroying it. It has become an important tool for the research scientist. It has been the basis of exacting techniques used in such recent discoveries as the uncovering of parts of the genetic code and the determination of its structure.

The first type of chromatography to be used was paper chromatography, a crude form of which will be done in this exercise. The mixture to be studied is first permitted to be adsorbed as a spot on the filter paper. A selected solvent then passes through the spot and ascends the paper by capillary action. As the solvent passes through the adsorbed mixture, some substances dissolve more rapidly than others, and thus travel along with the solvent at a faster rate. They thus travel farther up the paper before they are adsorbed by the paper again at a higher level. When the paper is removed from the solvent, the various components of the mixture may be present as spots on different heights. If you already know how high the various known substances ascend in this solvent system, you may identify the unknown substances in the mixture. Descending and circular methods of paper chromatography are also employed.

You can measure the heights of each of the spots from the point of application to the level to which it has moved within a specified time. This value is called the R_f value.

$$R_f = \frac{\text{height of the spot}}{\text{height of solvent front}}$$

There are several other types of chromatography in use today such as Thin-Layer Chromatography (TLC) which you will be hearing about later on.

MATERIALS NEEDED: A piece of filter paper one inch wide and about eight inches long, a glass rod, stapler, beaker, mixture of food colors.

PROCEDURE:

1. Put one end of the strip of filter paper around a glass rod and staple it in place.
2. Make a mixture of food colors by placing several drops of each of the food colors into a beaker and then mix thoroughly.
3. Add a little water to the mixture to dilute slightly.
4. Lay the glass rod across the top of the mixture in such a manner that the strip of paper is just below the surface of the mixture.
5. Watch what happens. Within minutes, the solution will rise on the paper and gradually each color that you put into the mixture will separate and form a separate band across the width of the strip. Every color that you place in the mixture will be seen on the filter paper as a separate band and in a different location.

Thus you see that with this very crude method of chromatography you were able to separate the components of a mixture. The more refined methods will be used to separate other mixtures as you go along.

THIN-LAYER CHROMATOGRAPHY

AIM: To become familiar with the technique of thin-layer chromatography.

As was previously mentioned chromatography is a rapid method of separating and identifying the components of a mixture without destroying them. In paper chromatography, the mixture to be studied is first permitted to adsorb as a spot on the filter paper.

We will use a new chromatographic procedure known as Thin-Layer

Chromatography (TLC). Instead of paper, the adsorbing surface is a layer of Silica Gel G (Merck) on a microscope slide. For the research worker, TLC is superior to paper chromatography in certain kinds of analysis. For us the advantage is speed.

There are prepared chromatogram sheets for TLC but in most cases the material has to be activated before using, whereas this method of making your own slides does not require much preparation.

We shall only give you in this exercise the method of preparing the microchromatic plates. The various materials that can be used with these slides will be forthcoming at a later time.

MATERIALS NEEDED: Slides, Coplin jar, Silica Gel G (Merck), chloroform, and methanol.

PROCEDURE:

1. Place 21 grams of Silica Gel G powder into a container such as a Coplin jar and then add a mixture of 33.5 ml chloroform and 16.5 ml of methanol. Stir vigorously for several minutes to form a slurry. When not in use, the jar should be covered with a greased top to prevent evaporation. The slurry may become thick because of evaporation and additional solvent may be added to keep the slurry at the proper consistency.

2. To dip the slides, new slides are not necessary as long as the slides have been washed and cleaned in the usual manner in any detergent and then air-dried. Place 2 slides together back-to-back so that one extends slightly over the other at the top and insert both in the slurry to the depth of about 1 cm from the top of the slides. Hold the slides at the top while doing this. Withdraw the pair of slides, scrape the bottom edge of the slides on the rim of the Coplin jar to remove excess silica gels. Allow to air dry for a few seconds.

3. Holding the slides firmly, pull the upper slide away from the lower slide and place them, silica side up, on a paper towel. Since these slides are prepared with a non-aqueous solvent, they are ready for use at once without further drying in an oven. After the slurry has been mixed, enough microchromatic plates can be prepared for an entire class in a few minutes.

4. Coated slides can be used after 2 weeks if left undisturbed. Jarring them will cause the dried gel to disintegrate.

NOTE TO TEACHER: There will be enough slides prepared for use by a class of 20 students.

UNIT II

CHEMISTRY OF LIFE

A. BASIC TERMINOLOGY

1. Atoms are composed of protons, neutrons, and electrons.
2. A particular element is composed of only one type of atom.
3. Compounds are composed of two or more types of atoms chemically united.
4. Organic compounds differ from inorganic compounds in that the former always contain both carbon and hydrogen.
5. Mixtures are composed of two or more types of atoms and/or compounds not chemically united.
6. Bonding is the sharing of a pair of electrons between the atoms of an organic compound. This bond holds these atoms together.

    a. When writing a structural formula, a single line represents a bond (a pair of shared electrons). A double line represents two pairs of shared electrons.

    b. The making and breaking of bonds involves an exchange of energy.

B. CHEMICAL ELEMENTS IN LIVING MATTER

The cell is a complex "chemical factory." It is composed of some of the same elements found in the nonliving environment: carbon, hydrogen, oxygen, nitrogen, sulfur, phosphorus, and traces of magnesium, iodine, iron and calcium.

C. CHEMICAL COMPOUNDS IN LIVING MATTER: Organisms consist of inorganic and organic compounds.

1. *Inorganic compounds*: The principal inorganic compounds are water, salts, inorganic acids, and bases. Water is very important as a solvent for other compounds.

2. *Organic compounds*: Organic compounds are unique to life and are always part of, or a product of, living systems. The principal organic compounds include carbohydrates, lipids, proteins, and nucleic acids.

a. *Carbohydrates*

(1) Carbohydrates consist of the elements carbon, hydrogen and oxygen. Hydrogen and oxygen are present in a 2:1 ratio.

(2) Structure: The principal carbohydrates include sugars and starches. The names of carbohydrates typically end in -ose, e.g., gluc*ose*, amyl*ose*, sucr*ose*, fruct*ose*, and so forth.

In general, sugars are low-molecular-weight compounds which are sweet and water-soluble. Included among the sugars are monosaccharides such as glucose, fructose, and galactose, and disaccharides such as maltose, lactose and sucrose. The structural formulas of a representative monosaccharide, glucose, and a disaccharide, maltose are:

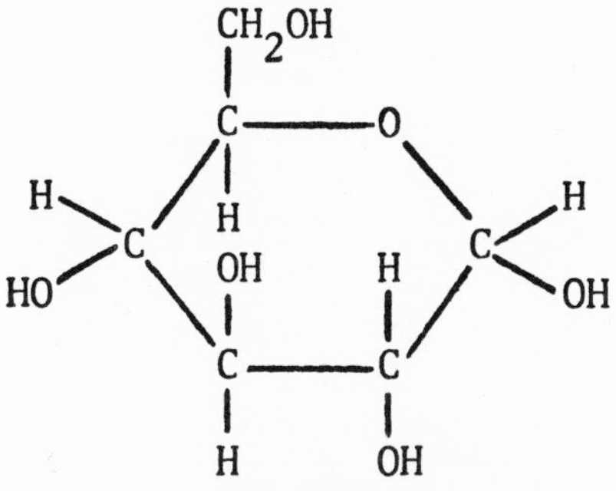

Glucose

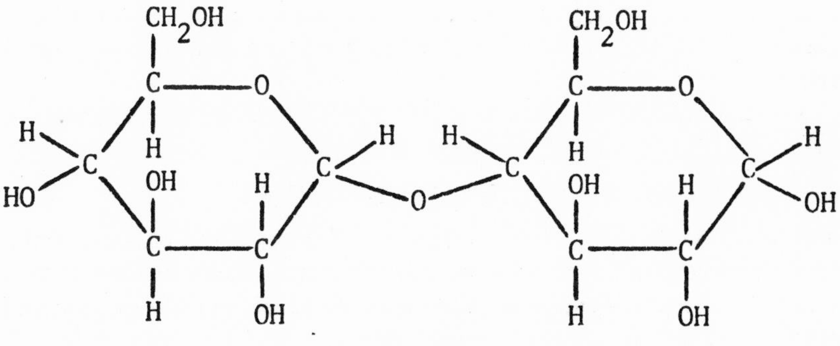

Maltose

Diagram 3.

Glucose molecules may combine to form maltose by dehydration synthesis. When maltose is formed from two glucose molecules a water molecule is released.

In general, starch is a high-molecular-weight compound which is slightly soluble in water. Starch is a complex substance consisting of amylose and amylopectin. The polymer concept is applicable where very large molecules are made up of repeating units of one type. It is also found in proteins and nucleic acids.

Amylose is a polysaccharide consisting of hundreds of repeating glucose units bonded together in a straight chain. It is slightly soluble in water.

Amylopectin also consists of hundreds of glucose units which are, however, bonded together to form a large branching molecule. It is insoluble in water. Symbolically, the molecules may be represented as follows:

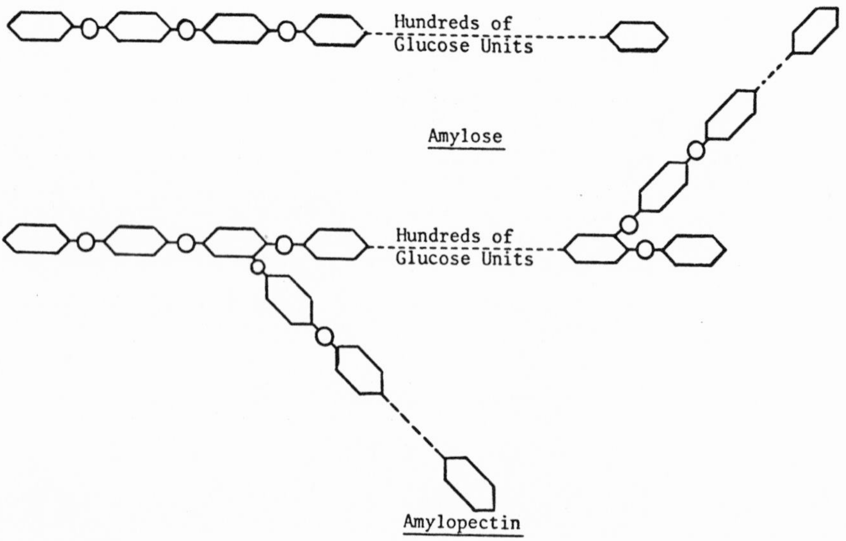

Diagram 4.

Cellulose, another widely occurring polysaccharide, is also a straight-chain molecule made up of repeating glucose units. It differs from amylose in the manner in which the glucose units are bonded together.

b. *Lipids*:

(1) Composition: Lipids, like carbohydrates consist of carbon, hydrogen, and oxygen. However, there is proportionately less oxygen in a

lipid molecule. The ratio of hydrogen to oxygen is always much greater than 2:1, and is not constant from one lipid to another. Lipids include fats, oils, and waxes.

(2) Structure: Generally lipids are small molecules and, unlike the other abundant organic compounds in living matter, are not polymers. Typically, lipids consist of 3 fatty acids bonded to a molecule of glycerol. Glycerol is a 3-carbon alcohol with the following structural formula:

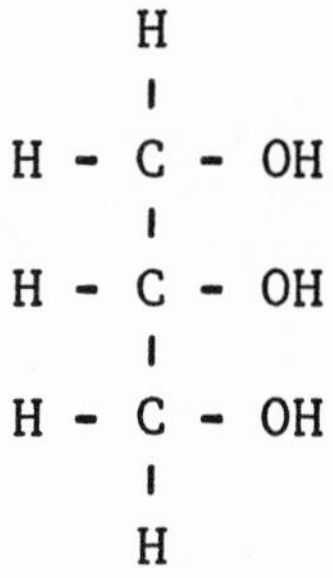

Diagram 5.

Fatty acids are long-chain hydrocarbon molecules which contain the characteristic organic acid group (-COOH) giving acid properties to the compounds. A general fatty acid structural formula appears as follows:

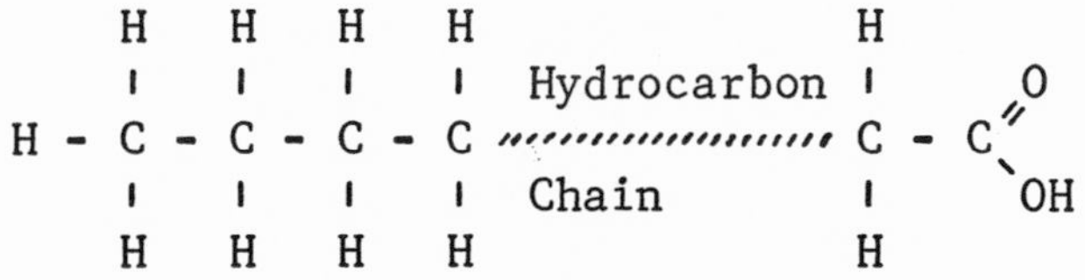

Diagram 6.

Characteristic groupings of atoms which show up frequently in different compounds are often referred to as groups, or, sometimes, side-groups. Examples include: $-CH_3$=methyl group; -COOH=carboxyl (acid) group; -OH=hydroxyl group.

The characteristic lipid structure is indicated by the structural formula shown in Diagram 7.

c. *Proteins*:

(1) Composition: Proteins contain carbon, hydrogen, oxygen,

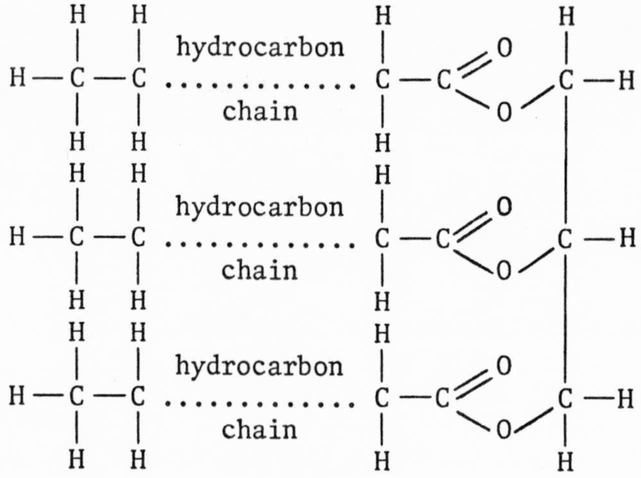

Diagram 7.

nitrogen, and, in many instances, sulfur.

(2) Structure: Proteins are high-molecular-weight polymers made up of repeating amino acid units. Although dozens of different amino acids have been identified, only 20 are typically found in living systems.

Amino acids have the general formula shown in Diagram 8 and derive their name from the amino (NH_2) group and acid (COOH) group which are characteristic of the compounds.

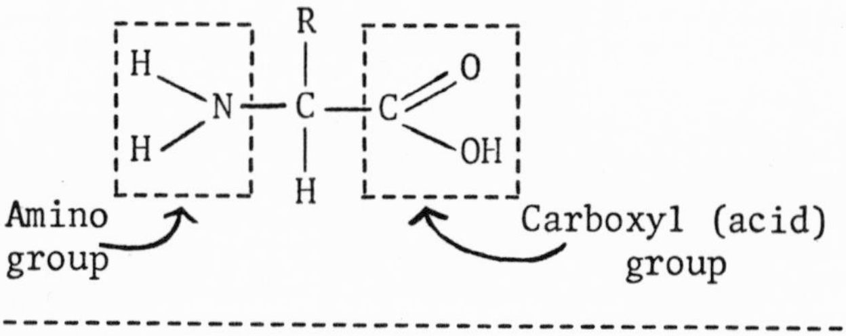

Diagram 8.

R represents a variable side-group and is the basis for the variety of different amino acids. When R is a methyl group, for instance, the amino acid is alanine.

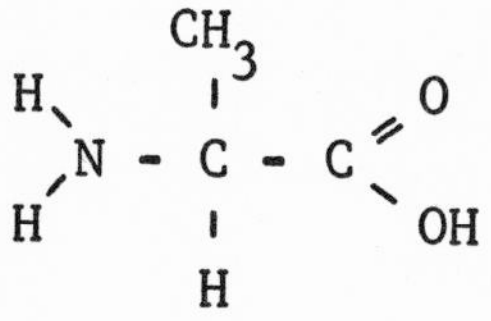

Diagram 9.

When amino acids bond to each other to form proteins, a C-N bond forms between the carboxyl group of one amino acid and the amino acid of an adjacent amino acid. The C-N bond is known as a peptide bond, and a chain of amino acids thus bonded is known as a polypeptide.

There is an extremely large (potentially infinite) number of different proteins. The bases for variability include:

(a) difference in the number and kinds of amino acids present.

(b) differences in the configuration of protein molecules due to attraction between different parts of the polypeptide chain producing a variety of shapes—straight, chain, globules, helices, and so forth.

d. *Nucleic Acids*:

(1) Composition: Nucleic acids contain phosphorus in addition to carbon, hydrogen, oxygen, and nitrogen.

(2) Structure: Nucleic acids are the largest organic molecules known. They are high-molecular-weight polymers made up of repeating units known as nucleotides.

Nucleotides, although unit molecules of the nucleic acids, are relatively complex molecules consisting of 3 sub-units: phosphate, sugar, and nitrogenous base. Their basic organization may be presented

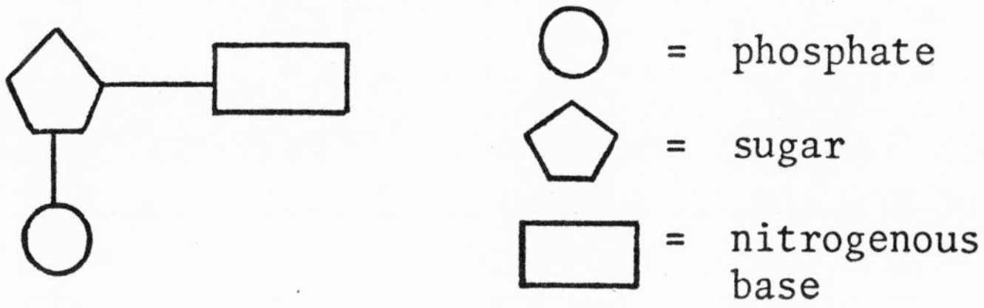

Diagram 10.

The sugar is always either ribose or deoxyribose. Ribose is found in ribonucleic acids and deoxyribose in deoxyribonucleic acids.

D. CHEMICAL ACTIVITY IN LIVING MATTER:

Living matter is constantly in a state of dynamic chemical activity. Perhaps the most significant distinction between living and nonliving matter is the continuous, controlled chemical activity in living systems. For the sake of simplicity it is suggested that all enzyme names end with the suffix "-ase" (e.g., an enzyme hydrolyzing protein would be called "protease").

1. *The role of Enzymes*: The organic catalysts called enzymes are the principal means by which chemical activity in a living system is regulated.

Each chemical reaction occurring in an organism requires an enzyme.

Enzymes modify, by either accelerating or suppressing, the rate of reaction.

a. Structure: There are thousands of different enzymes in cells. The great variety of enzymes is due to their protein nature.

(1) Protein nature: All enzymes are either exclusively proteins or are proteins with attached, nonprotein, side-groups known as coenzymes. Often, vitamins function as coenzymes. When enzymes have both a protein and a nonprotein component, the protein part alone will not function in the absence of the coenzyme.

(2) Usually enzyme molecules are enormously larger than the molecules with which they interact. This observation, coupled with other data, has lead to the assumption that only a small portion of the enzyme molecules functions in enzyme action. This localized region is called the active site of the enzyme.

b. Function: Little is known about the manner by which enzymes achieve their catalytic control over cell reactions. However, sufficient evidence has accumulated to permit biochemists to develop a model of enzyme action which is useful in visualizing the nature of their function and is consistent with the available evidence.

(1) Enzyme-substrate complex: It is thought that, for an enzyme to affect the rate of a reaction, the following steps take place:

It must form an association with the substance or substances whose reaction rate it affects. These substances are known generally as substrates.

This association between enzyme and substrate is thought to involve a close physical association between the molecules, but does not involve bond formation between them.

The association between enzyme and substrate is known as enzyme-substrate complex.

It is while the enzyme-substrate complex is formed that enzyme action takes place.

Upon completion of the reaction, enzyme and products separate, and the enzyme molecule is then available to form additional complexes.

Enzymes, like all molecules (particularly large molecules), are subject to destructive forces which tend to make them disintegrate in time. Thus, a constant supply is needed by all cells.

(2) "Lock-and-key" model: The fact that a particular enzyme will usually only interact with a single type of substrate molecule has given rise to the "lock-and-key" model of enzyme specificity. Like a key that will only open a particular lock, a particular enzyme will usually only form a complex with one particular type of substrate molecule.

Specificity varies to some extent. In some instances an enzyme is highly specific for a single compound. In other instances enzymes exhibit group specificity and will form complexes with a number of related compounds.

c. Factors Influencing Enzyme Action: The rate of enzyme is not fixed, but varies according to conditions in the environment of the reacting substances. Such factors as pH, temperature, and relative amounts of enzyme and substrate determine the rate of the enzyme action.

(1) pH: The rate in which enzyme-regulated reactions occur varies according to the hydrogen ion (H^+) concentration in the environment.

Hydrogen ion concentration is usually indicated by means of the pH scale.

The pH scale extends from 0–14. A pH of 7 is neutral. Acids have a pH less than 7 and bases have a pH greater than 7.

For many enzyme-controlled reactions a pH of 7 provides an environment for optimal rate of enzyme reaction. pH greater or less than 7 would tend to slow down the reaction rate. This situation is summarized in Diagram 11.

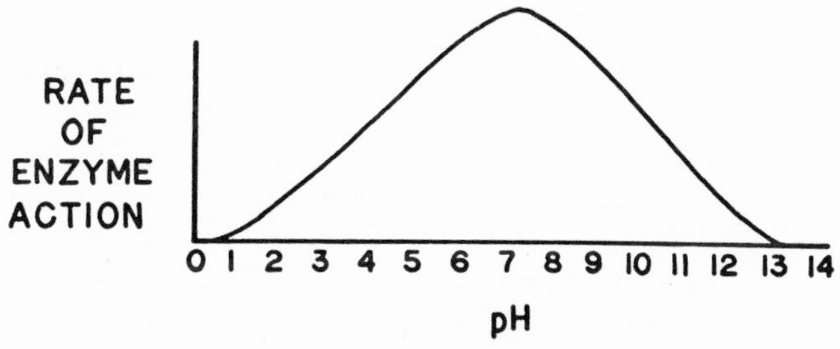

Diagram 11.

(2) Temperature: At low temperatures enzyme-controlled reactions tend to behave the same as uncatalyzed reactions. In general, as temperature increases, the rate of enzyme action increases.

At relatively high temperatures, however, the "shape" of enzyme molecules tends to be altered, thus rendering the enzyme ineffective. This distortion of enzyme molecules at high temperature is known as enzyme deactivation. For many enzymes deactivation occurs at around 40°C.

The response to changing temperatures for many enzymes is illustrated in Diagram 12.

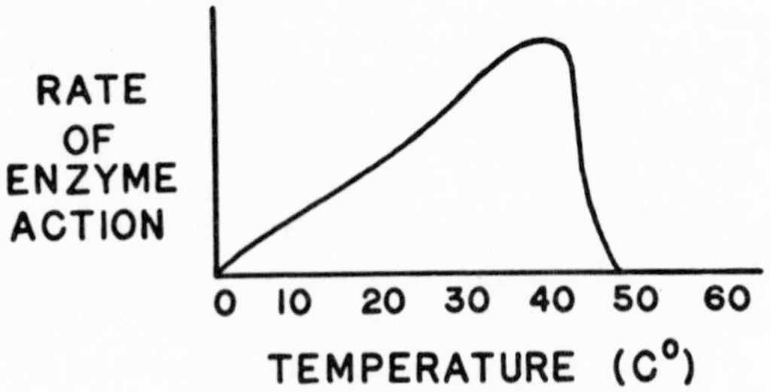

Diagram 12.

(3) Relative Amounts of Enzyme and Substrate: The rate of enzyme action also varies according to the amount of available substrate molecules. When an excess of enzyme is added to a system with a fixed concentration of substrate, the rate of enzyme action tends to increase to a point and then remain fixed as long as the substrate concentration remains constant.

Diagram 13 illustrates the pattern of enzyme action rates when an excess of enzyme is added to a system with a fixed substrate concentration.

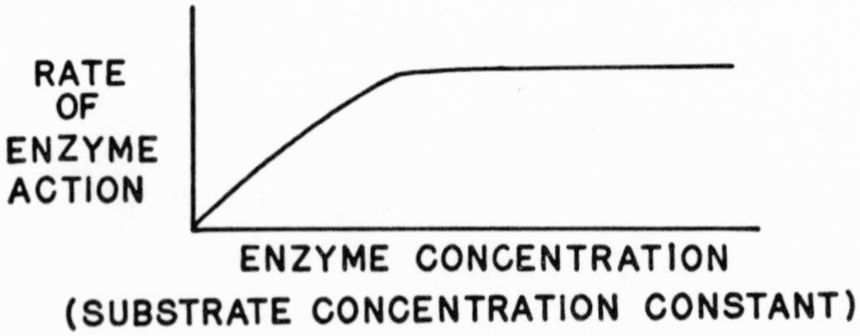

Diagram 13.

Conversely, if the level of enzyme concentration is constant, the rate of enzyme action will increase as substrate is added to the system but will stabilize at the point where all available enzyme molecules are activity involved in the reaction.

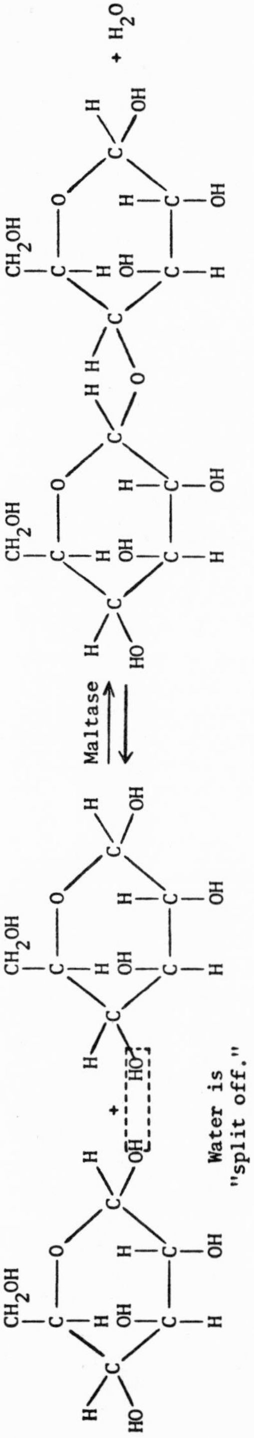

Diagram 14.

2. *Reactions*: Although there are countless reactions taking place in cells at all times, hydrolysis and dehydration synthesis account for much of the biochemical activity in living systems.

a. Dehydration synthesis: The process by which large organic molecules are built up from their building blocks with the release of water molecules.

b. Hydrolysis: The process by which these same molecules are broken down into their building blocks with the addition of water molecules.

Both hydrolysis and dehydration synthesis are enzyme-controlled, reversible reaction.

Diagram 14 gives an example of the dehydration-synthesis–hydrolysis reactions associated with glucose and maltose.

MIXTURES AND COMPOUNDS

AIM: To study the difference between elements, mixtures and compounds.

The universe is made up of about 100 fundamental substances called elements. The smallest part of an element that will combine with other elements is called an atom. Generally, atoms will combine with atoms to form molecules. The elements consist of molecules, the smallest part of an element to retain the properties of the element. Identical atoms are found in the molecules of elements. When two or more unlike atoms combine, a compound is formed. The molecules of the compounds keep the properties of the compound and consist of unlike atoms. There are so many different substances found in the world because of the tremendous number of chemical combinations possible.

MATERIALS NEEDED: Powdered sulfur, iron filings, carbon disulfide, filter paper, test tubes, test tube holder, magnet, funnel, watch glass, Bunsen burner, ring stand, ring, 250 ml beaker, balance, cork stoppers.

PROCEDURE:

1. Take a small sheet of paper and spread 1/4 teaspoonful of iron filing on it. Bring a magnet to the filings and the magnet will be attracted. Now place 1/4 teaspoonful of sulfur on a sheet of paper and try with the magnet. It will not be attracted.

2. Into a test tube, place a pinch of iron filings and into another a pinch of sulfur. Add carbon disulfide to both to the depth of one inch. Stopper each tube and shake gently. The sulfur dissolves in the carbon disulfide while the filings do not. (Be certain there are no flames near the carbon disulfide.)

3. Now mix together a little of the filings and sulfur on a sheet of paper.

When you apply the magnet to it, the magnet will attract the iron and the sulfur will remain on the paper.

4. Mix the filings and the sulfur, and place into a test tube with about 2 inches of carbon disulfide. Stopper the tube and shake. Now filter the liquid by putting the funnel into the ring on the ring stand and filtering into a beaker. Place a few drops of the filtrate on a watch glass and let it evaporate. Within 10 minutes, you will find a residue left. The iron filings will be left on the filter paper in the funnel while the sulfur crystals appear as the carbon disulfide evaporates.

5. Weigh out 7 grams of iron filings and 4 grams of powdered sulfur and mix on a square of paper. Place into a test tube and heat until you see the two substances mixing. Cool by placing the end of the tube in cold water. The tube will break. Use the magnet on the black substance of the tube. Nothing will happen. (Caution: Make certain that no carbon disulfide is anywhere near an open flame.)

CONCLUSIONS:

Mixtures can be made in any proportions while compounds can only be formed in definite proportions of the elements in them.

You also saw that for the iron and sulfur to unite chemically it was necessary for the mixture to be heated and then it did not show the same characteristics as the mixture. The individual elements of the mixture were recovered through filtration and evaporation.

THE MEASUREMENT OF pH*

PURPOSE: To observe the effect of dilution upon pH.

MATERIALS NEEDED: For each group of two students: 9 baby-food jars, 0.01M HCl, medicine dropper, 0.01M NaOH, stirring rod, pH paper strips, paper toweling, pH color chart, various biological solutions, distilled water.

INTRODUCTION: Water ionizes as follows: HOH into $H^+ + OH^-$. The number of hydrogen (H^+) ions is equal to the number of hydroxide (OH^-) ions, and the water is *neutral*. If the number (or concentration) of H^+ ions in a solution is greater than the number of OH^- ions, the solution is *acid*. If the number of OH^- ions is greater than the number of H^+ ions, the solution is *basic* (or alkaline).

The following are examples of the ionization of acids and bases:

* Starred experiments throughout the book should be checked for materials at least two days in advance. Some materials must be prepared ahead of time.

Acids

$HCl \rightarrow H^+ + Cl^-$ (chloride ion)
$HNO_3 \rightarrow H^+ + NO_3^-$ (nitrate ion)
$H_2SO_4 \rightarrow 2H^+ + SO_4^=$ (sulfate ion)

Bases

$KOH \rightarrow K^+ + OH^-$
$NaOH \rightarrow Na^+ + OH^-$
$Ca(OH)_2 \rightarrow Ca^{++} + 2OH^-$

Neutralization is the process in which a base and an acid unite to form water and a salt.

BASE	+	ACID $\rightarrow$ WATER	+	A SALT
KOH	+	$HCl \rightarrow HOH$	+	KCl
$NaOH$	+	$HCl \rightarrow HOH$	+	$NaCl$
$Ca(OH)_2$	+	$H_2SO_4 \rightarrow 2HOH$	+	$CaSO_4$

When neutralization has been properly performed, there is an excess of neither H^+ ions nor OH^- ions, as in water.

A *Salt* may be regarded as a compound formed by the union of the positive portion of a base and the negative portion of an acid. Table salt ($NaCl$) is only one example of the many kinds of salt.

pH Scale: Scientists use the pH scale to indicate the relative concentration of H^+ ions and OH^- ions in a solution. The scale runs from 1 to 14. The midpoint, 7, indicates a neutral solution in which the concentration of H^+ ions and OH^- ions are equal. Proceeding toward the lower numbers indicates greater acidity; proceeding toward the high numbers indicates greater alkalinity. One of the purposes of this exercise is to discover how great a change in concentration of the ions is represented by a change of one unit on the pH scale. The pH of solutions is of great importance in the chemical reactions of the living cell.

Indicators: The pH value of a solution can be measured accurately by means of an electrical device called a pH meter. The pH can also be measured, but less accurately, by means of complex dyes which change colors as various pH values. Such chemicals are called *indicators*. One such indicator which is familiar to many students is *litmus*, which is pink in acid solution and blue in basic solution. In this exercise you will use *pH paper*. This is paper which has been treated with a mixture of indicators in such manner that the color produced on the paper can be compared with a reference color scale to show the pH.

CAUTION: The acid and base furnished to you are quite dilute and are, therefore, not harmful. However, avoid getting any into your eyes or mouth or on your skin. In case of accident flush at once with large amounts of water and notify your instructor.

NOTE: The general procedure in this experiment is to make a series of dilutions of an acid and of a base, and to determine the pH of each dilution. This is followed by the neutralization of the acid and base.

MDF: The usual unit of measurement of volume used in biological laboratories is the *milliliter* (ml) which is one-thousandth of a liter. Because it is impractical to sterilize the large numbers of *pipettes* needed for many classes to measure volume in milliliters, we shall use medicine droppers and shall invent for measuring volume—*the medicine-dropper-full*, abbreviated *MDF*.

Serial dilutions: In order to prepare a series of dilutions, each of which is 1/10 the concentration of the previous solution, the procedure diagrammed below will be employed.

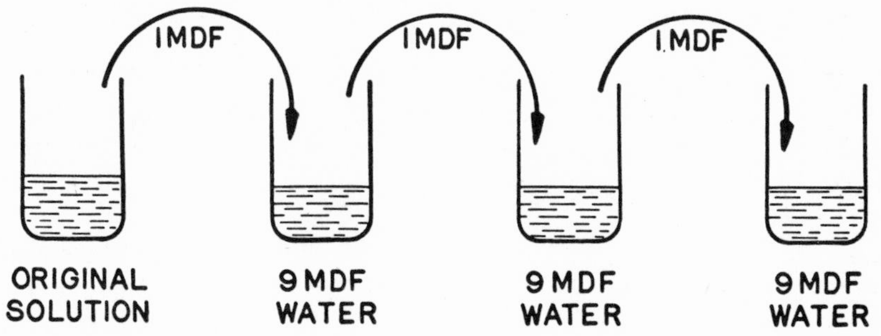

Diagram 15.

Use of the Medicine Dropper: Although the medicine dropper seems to be a simple piece of laboratory apparatus, several precautions should be observed for its proper use:

1. The liquid in the dropper should never be allowed to run into the rubber bulb. The liquid may be a substance which reacts with the rubber. It is also more difficult to clean the rubber bulb than the glass portion of the dropper. Always hold the medicine dropper in a vertical position with the rubber bulb at the top. Air pressure prevents the liquid from running out the bottom.

2. In withdrawing fluid, it sometimes makes a difference whether the rubber bulb of the medicine dropper is squeezed *before* the dropper is inserted into the liquid or *after* it is inserted. Squeezing the bulb after insertion causes agitation of the liquid as air is introduced. In some cases, as when you are working with a protozoan culture, this agitation may be undesirable. In some cases, the introduction of air may cause oxidation of substances in the liquid. In general, do not agitate the liquid by introducing air unless there is a reason to do so.

3. Fairly uniform MDF's may be measured by squeezing the rubber bulb *tightly* each time, before insertion into the liquid. The same medicine dropper should be used for any single series of measurements. In experiments where you employ a medicine dropper to deliver a specified number of *drops*, also utilize the same medicine dropper throughout.

PROCEDURE:

A. *Acid Dilutions*: (Wash your medicine dropper thoroughly before starting)

1. The starting solution is hydrochloric acid at a concentration known to the chemist as 0.01 M (1/100 Molar). Introduce the acid into Jar A1 to a depth of about 2.5 cm (1 inch).

2. Into each of Jars A2, A3, and A4 introduce 9MDF distilled water.

3. Remove MDF of solution from Jar A1 and add it to the 9MDF in Jar A1. Take up this diluted mixture into the medicine dropper several times in order to wash out the dropper. The concentration of acid in Jar A2 is now 1/10 that of Jar A1, or 0.001M HCl.

4. Continue in like manner to make serial dilutions into Jars A3 and A4.

5. Indicate the concentrations on the Table.

B. *Basic Dilutions*: (Wash your medicine dropper thoroughly before starting)

1. The starting solution is 0.01M NaOH (sodium hydroxide). Introduce this solution into Jar B1 to a depth of about 2.5 cm.

2. Follow the procedure outlined above to make serial dilutions into Jars B2, B3, and B4. Indicate concentrations on the Table.

C. *Determination of pH*: Place a strip of pH paper on a piece of notebook paper. Using the stirring rod, transfer a drop from Jar A1 to the strip of paper and determine the pH by comparison with the color which appears with the color chart. Enter the pH on Table I. Do the same for the other jars.

Before each determination, wash the stirring rod and dry it on paper toweling. Continue to determine and record the pH of each solution.

D. *Neutralization*: An additional jar has been furnished to you. Take 5MDF of the original acid solution and then 5MDF of the original basic solution and place into jar. Stir with a clean stirring rod and test the pH. It should come out neutral.

OBSERVATIONS:

The pH of BIOLOGICAL SUBSTANCES: Determine the pH of substances such as:

lemon juice . . . .
urine . . . .

blood plasma . . . .
vinegar . . . .
milk . . . .
raw egg . . . .
bicarbonate of soda . . .
soda pop . . . .
soil . . . .
solution of bile salts . . .
saliva . . . .
artificial gastric juice . . .

JAR	CONCENTRATION	pH
Acids A-1	0.01M HCl	
A-2		
A-3		
A-4		
Base B-1	0.01M NaOH	
B-2		
B-3		
B-4		

QUESTIONS:

1. What is an acid?
2. What is a base?
3. Give the equations for the ionization of 3 acids and 3 bases.
4. Define neutralization.
5. Write the equation for the neutralization of NaOH and HCl.
6. What is salt?
7. What is the pH of pure water?
8. Which is more acid—a solution of pH 5 or pH 2?
9. Which is more alkaline—a solution of pH 7.2 or pH 7.4?
10. What indicator will be used in this exercise?
11. As the acid is diluted, what kind of change takes place in the pH?
12. As the base is diluted, what kind of change takes place in the pH?
13. What change in concentration of the H^+ and OH^- ions accompanies a change in *one* pH value?
14. Explain any discrepancy between the results you obtained and the theoretical results.

Acknowledgments: This laboratory exercise is one that is performed at the Far Rockaway High School, Far Rockaway, N.Y., and Mr. David

Kraus, Chairman of the Science Department, has devised the ingenious method of using medicine droppers instead of pipettes and calling the measurement MDF.

ENZYMES*

AIM: To acquaint one with the workings of two enzymes: urease and catalase.

Enzymes are organic catalysts permitting the organism to carry on reactions rapidly and under the "mild" conditions of the cell. This exercise will deal with two intra-cellular enzymes.

MATERIALS NEEDED: For Each Group of Students: Rack with 6 test tubes; urease solution in dropping bottle; red litmus paper; 3% hydrogen peroxide; water bath; Bunsen burner; tripod; yeast; liver (freshly ground with sand); labels; urea, saturated solution; potato cubes, about 5 mm on each side; wood splints; beaker of blood with medicine dropper; hydrogen peroxide, 20%.

PROCEDURE:

I. Urease

Urease catalyzes the breakdown of urea, $CO(NH_2)_2$ to *carbon dioxide* and *ammonia*. The student is to make a rough comparison of the temperature needed to break down urea with, and without, this enzyme. Ammonia may be detected by its odor and by its effect on wet red litmus paper.

Introduce about 2 inches of urea solution into each of two test tubes.

Tube #1 (heat): Fold a strip of wet red litmus and place in the mouth of the tube. Label the tube and place in a water bath.

Tube #2 (Enzyme): Add two drops of urease solution and shake. Place a strip of wet litmus paper in the mouth of the tube. Keep this tube at body temperature by holding in the hand.

At 5-minute intervals examine the litmus paper in both tubes. The tube #2 will show some results as an odor of ammonia will be given off and the red litmus paper will turn blue.

II. Catalase:

It is believed that this enzyme prevents the accumulation of poisonous peroxides which might otherwise form during cellular respiration.

1. Place a pinch of yeast into a test tube and add a little of the 20% hydrogen peroxide. There will be great frothing. Light a wood splint and then use the glowing splint to show that oxygen is being given off as it lights up again when thrust into the froth.

2. Now into each of the other 3 test tubes place a drop of blood, a cube

of potato and a small amount of fresh liver, respectively. To each test tube add 2 inches of 3% hydrogen peroxide. If there is a vigorous production of gas, try to identify the gas by means of a glowing wood splint.

QUESTIONS:

1. Urea in the presence of urease breaks down into . . . . . . and . . . . . .
2. What indication do you have that catalase is widely distributed in living things?
3. Why is it that "peroxide" placed on your skin does not bubble, but that it does bubble if you have a cut?
4. How does peroxide kill microorganisms in a cut?
5. Can you work up a controlled experiment to determine the effect of boiling upon catalase from liver?

This exercise is adopted from one done in Far Rockaway High School, Far Rockaway, N.Y.

MORE ABOUT ENZYMES*

AIM: To demonstrate that enzymes are used for two very important processes: hydrolysis and synthesis.

Some of the most important substances in the cell from the standpoint of cellular activities or metabolism are enzymes. We usually think of enzymes only in connection with the digestive system. Many specific digestive enzymes exist, but these act outside the cell. These enzymes are but a very small fraction of the total number and kinds of enzymes produced within living cells. Enzymes are really proteins and are responsible for the many chemical reactions taking place in the living cell.

Enzymes aid and speed up chemical reactions, but are then recovered intact when the reaction is completed. Thus, only a minute number of molecules of a particular enzyme can be responsible for many reactions.

Some chemical reactions in the cell result in the synthesis of substances necessary for the life of the cell while there are other chemical reactions which result in the hydrolysis or breakdown of complex substances to simple ones.

In this exercise, we shall give you an example of both processes.

MATERIALS NEEDED: Agar, water, glucose-l-phosphate, petri dishes, small potato, mortar and pestle, clean sand, centrifuge, beakers, iodine solution, corn grains, cheesecloth, commercial corn starch, razor blade, Benedict's solution.

PROCEDURE:

(A) Synthesis by Enzymes: This process can be demonstrated in the school laboratory by allowing a potato extract, which contains the proper enzymes to change a simple sugar compound, such as glucose-l-phosphate, to a starch, amylose.

1. Prepare a medium of 1% agar in water. Heat to melt the agar. Then add glucose-l-phosphate to make a 0.5% solution. When completely dissolved pour into petri dishes and allow to harden.

2. Meanwhile prepare the potato extract enzymes as follows: Peel and cut up a small potato. Grind to a pulp with a mortar and pestle along with a little clean sand to make it easier. Centrifuge the pulp until the supernatant is clear. Decant the supernatant into a beaker.

3. Test a small sample of the extract with the iodine solution to make certain that there is no starch present. If there is a blue-black color, centrifuge until the extract is starch free and gives no reaction with the iodine.

4. Place approximately 6 drops, well spaced from each other, on the agar plate and test at 5 minute intervals by adding one drop of iodine to one drop of the potato extract. You will soon find that there is a positive starch test proving that a simple sugar can be changed into a starch.

(B) Hydrolysis by Enzymes: There are chemical reactions which result in the hydrolysis, or breakdown, of complex substances to simple ones, such as the conversion of starch to sugar.

This process can be demonstrated in the laboratory by observing the action of enzymes from corn scutellum (cotyledon) on a starch mixture.

1. Place several grains of corn in a cheesecloth bag and soak in running tap water for 24 hours.

2. Prepare a test medium containing 1% agar and 1.5% soluble commercial corn starch. Heat the mixture to boiling, pour into petri dishes and allow to harden.

3. Use a razor blade to remove the endosperm from the corn grain. Discard the endosperm and place the cut surface of the scutellum of the corn embryos on the hardened agar in the petri dishes. Place two embryos on each petri plate.

4. For the next 24 to 48 hours leave the plate at room temperatures in a dark spot. Then test the agar with iodine solution for the presence of starch. There will be a gradual lessening in the amount of starch found.

5. At the same time test the agar at the site of each embryo with Benedict's solution for the presence of reducing sugar. There will be an increase in the amount of sugar found indicated by a deepening of the color change with the Benedict's solution.

CONCLUSIONS:

1. Why is iodine used in Part A of the procedure?
2. Why is iodine used in Part B of the procedure?
3. Can Benedict's solution be used for all types of sugar?

CARBOHYDRATES*

AIM: To study the composition and means of identification of all types of carbohydrates.

Carbohydrates are compounds usually made up of carbon, hydrogen, and oxygen in which hydrogen and oxygen occur in the proportion of 2 to 1 so that the basic pattern is CH_2O. Cellulose which is the most abundant organic material in nature and starch which is an important foodstuff for man belong to this class of compounds.

Simple carbohydrates that contain six basic units are known as monosaccharides, examples of which are glucose and fructose. A disaccharide is formed when two monosaccharides unite with a loss of a molecule of water. Sucrose and lactose are examples of disaccharides.

Starch is a polysaccharide as is cellulose and glycogen (animal starch). These are substances of large molecular weight, in the order of millions, that contain thousands of glucose units linked together in a continuous chain.

When the poly- or disaccharides are heated in solution in the presence of strong acid, they are split up into the component sugars, the monosaccharides, which are the simple sugars. Thus cellulose, starch and glycogen yield glucose; sucrose yields equal parts of glucose and fructose while lactose yields equal parts of glucose and glactose.

In this exercise we shall do various types of tests on the various types of carbohydrates.

BENEDICT'S IDENTIFICATION TEST:

Materials Needed: Test tubes, test tube rack, Bunsen burner, 1% solutions of glucose, fructose, sucrose, lactose. (The 1% solution is prepared by dissolving 10 grams of the sugar in a liter of distilled water.) Benedict's solution, graduated cylinder, water bath.

Procedure:

Certain sugars, called reducing sugars, can be detected by means of the Benedict test. A portion of the reducing sugar molecule reacts with the copper sulfate in the Benedict's solution converting the copper sulfate to copper oxide, a red compound.

1. Place 5 ml of Benedict's solution in each of 4 tubes. Add 8 drops of each of the sugar solutions into a different test tube. Shake the tube gently in order to mix the solution. Place the tubes in a boiling water bath for about 5–10 minutes. When the tubes are cool, you will find the glucose has completely reduced and there is a deep red color to the solution because it is a monosaccharide. The same will be true of the fructose. There will be no reaction in the sucrose and the lactose because they are both disaccharides and will not reduce.

POLYSACCHARIDE IDENTIFICATION:

Materials needed: IKI reagent (iodine dissolved in potassium iodide);

2% starch solution, 2% glycogen solution; potato starch diluted with 3 or 4 volumes of water, filter paper.

Procedure:

1. Put a drop of IKI on a piece of filter paper. It will turn violet-brown in color since there is cellulose in the paper.

2. Place 2 ml of 2% starch solution into a test tube, add a single drop of IKI reagent and the resulting color will be a deep blue-black. If you heat the tube, the color will fade but when cooled under the water tap, the color will reappear.

3. Place 2 ml of the potato starch diluted suspension in a test tube and add the IKI. The color will be blue-black but not as deep. It, too, will fade on heating and return on cooling under the running water.

4. Place 2 ml of 2% glycogen solution into a test tube and add a drop of the IKI. The resulting color will be a red-brown which also will fade on heating and return upon cooling.

Thus you can see that different types of polysaccharides give different results with the same reagent thus indicating that each polysaccharide has its particular structure.

HYDROLYSIS:

Materials needed: Potato starch, concentrated HCl, test tubes, water bath, medicine droppers, Benedict's solution, IKI reagent, test tube rack.

Procedure:

1. Put 5 ml of potato starch into a test tube, add 5 ml of water and shake thoroughly. Then add 3 ml of concentrated HCl. Stir and place the test tube into the boiling water bath.

2. After 60 seconds, take a drop of the solution from the tube and place into 2 test tubes. Then test each tube, one with Benedict's for sugar and one with IKI for starch. You will find a gradual change from blue-black of unhydrolyzed starch to the red-brown of partially hydrolyzed starch and then to the absence of any reaction. As for the Benedict's test: at first it will be negative and gradually work up to a positive test when there is a negative test with the IKI.

Thus you can see how the starch gradually changes into sugar in the presence of an acid.

CONCLUSIONS:

Answer the following:

1. Which carbohydrates reduce the copper ion?
2. Which compounds are the reducing sugars?
3. Why is there a negative reaction on the part of some sugars toward Benedict's solution?

4. Do you know what are the chemical differences between glycogen, cellulose, and starch?

5. What significance can you attach to the various color changes during starch hydrolysis?

THIN-LAYER CHROMATOGRAPHY OF SUGARS*

AIM: To learn how to identify and separate various sugars out of a mixture in solution.

We have seen how we can identify sugars in various substances. However, through thin-layer chromatography we have a most modern method of separating and identifying various types of sugars.

MATERIALS NEEDED: Coated microchromatic slides (preparation in Unit I), Coplin jars, lactose, glucose, xylose, ethyl acetate, isopropanol, water, asbestos sheet, hot plate, micropipettes, plastic ruler in centimeters, stoppered bottles, balance, graduate cylinders, 85% phosphoric acid, naphthoresorcinol, ethyl alcohol.

PROCEDURE:

1. Prepare as many microchromatic slides as needed. The preparation of these slides is explained in Unit I.

2. Now prepare 1% solutions in isopropanol of lactose, glucose, and xylose. Label lactose #1, glucose #2, and xylose #3.

3. Prepare the solvent by mixing together 55 ml ethyl acetate, 36 ml isopropanol and 12 ml of water.

4. Prepare a mixture of all three sugars by mixing together equal amounts of each of the sugars into another bottle, and label it #4.

5. Number four coplin jars from 1 to 4 and place about 2 ml of the developing solution into each of the chromatojars. Cover the jar for a few minutes so that the vapors fill the container.

6. Spot the slides by using a different sugar for each slide as well as the mixture. Thus you will have 4 different slides. Fill a pipette with a solution by capillary action. Touch only the liquid to the silica gel surface without breaking the surface. Arrange 20 drops adjacent to one another to form a line across the plate about 1 cm from the bottom of the slide. Allow the spots to dry thoroughly and then place lactose #1 in the coplin jar marked 1 and so on until you have all 4 slides in the proper chromatojars.

7. The solvent will rise, and when it approaches the top of the silica gel layer, remove each slide from the jar. All the slides should be kept in for the same period of time so that all conditions will be nearly identical.

8. Allow the slides to air dry completely and then spray with the visualizer made up of 10 ml 85% phosphoric acid, and 0.2% naphthoresorcinol in 100 ml ethyl alcohol.

9. When the visualizer has dried slightly, place the slides on an asbestos pad on a hot plate which is set at 200°C.

10. Within a few minutes each of the slides will begin to show color. Slide #1, lactose, will show a brownish color; slide #2, glucose, will show a reddish color, while slide #3, xylose, will show a greenish color. Slide #4, which is the mixture, will show 3 bands of color and when compared to the first three slides, it can be seen that lactose is the band closest to the bottom of the slide, glucose is the middle band, and xylose is the band that traveled the farthest.

CONCLUSION:

We thus have another method of identifying sugars and no matter where the above sugars are found, if the thin-layer chromatography done is of the type just completed, you should be able to identify any of the three sugars, lactose, glucose, or xylose.

LIPIDS*

AIM: To study the various characteristics of lipids and their means of identification and hydrolysis.

Lipids, or fats, make up a large variety of chemical materials that do not mix with water, but readily dissolve in solvents like ether, chloroform or alcohol.

Some lipids in the body associate with other lipids and with proteins to form membranes which act as a barrier between one compartment and another. However, not all lipids are structural components and many of them supply energy. The storage of energy is done in the form of fat droplets in adipose tissue cells. In fact, fat is the most concentrated form of energy that living systems have devised.

There are various kinds of lipids such as the dietary fats like butter and corn oil which are mainly fatty acid esters of glycerol, known as triglycerides; the insulating matter of brain and nerve cell sheaths which are mainly cholesterol; the protective waxes and hydrocarbons of the skin; the fat-soluble vitamins; and many others. But all prefer an organic solvent to water, and most of them contain long-chain fatty acids.

In this exercise, we shall do several identification tests and the hydrolysis of a lipid to fatty acid.

IDENTIFICATION TESTS:

Materials Needed: Corn oil, test tube, brown wrapping paper, Sudan IV.

Procedure:

1. The simplest way to identify fats is to make use of their well-known ability to produce translucent marks on paper. With a medicine dropper add a drop of corn oil to the corner of a sheet of brown wrapping paper or any unglazed paper. To the opposite corner add a drop of water. When the fluids have dried, hold the paper up to the light and you will see a spot of grease still on the paper whereas there is absolutely no spot where the water was.
2. To 3 ml of water in a test tube, add a drop of Sudan IV. Now add 1 ml of corn oil to the tube and shake thoroughly. You will find that the dye has been taken up by the drops of the oil.
3. Instead of Sudan IV, you can add Sudan III and you will see the oil droplets turning pink as they take up the dye.

HYDROLYSIS OF LIPID TO FATTY ACID:

Materials needed: Water, corn oil, soap solution, pancreatin, water bath, phenolphthalein, 3 test tubes, test tube rack, water bath.

Procedure:

1. Into test tube #1, place 10 ml of water, a few drops of corn oil, 2 ml of soap solution, 5 ml of pancreatin (pinch of pancreatin in 10 ml of water).
2. Into test tube #2, place 10 ml of water, a few drops corn oil, and 2 ml of soap solution.
3. Into test tube #3, place 10 ml of water, a few drops of corn oil, and 5 ml of pancreatin.

Add a few drops of the phenolphthalein indicator into all three tubes so that the solutions become light pink. Place all three tubes in a water bath. After some time tube number 1 will become colorless showing that the corn oil has changed into fatty acids.

CONCLUSION:

1. Lipids must have a combination of an emulsifier and pancreatin to break down fatty acids. What must be present to enable pancreatin to break down lipids into fatty acids? Why do you think it is necessary?
2. In what part of the human body are the fats changed into fatty acids?

PROTEINS*

AIM: To study the reactions of proteins to various tests used as a means of identification and also to learn something of the composition of proteins.

Proteins are nitrogen-containing compounds. Proteins are universally distributed in all cells of all organisms. They are associated with the structure and function of all living systems. More than half of the dry weight of all living tissues is made up of proteins.

Proteins show extreme variations in size and shape and their biochemical composition is more varied than most other biological materials. The smallest of the proteins have molecular weights of 6,000 while there are others which range in size into the millions. Proteins which combine with nucleic acids are the largest and most complex substances in cells.

The enzymes which make possible the thousands of reactions of metabolism are all proteins. Many hormones are proteins as are the toxins given off by bacteria within a host. Antigens, silk, wool, hair, horns, and hoofs are all proteins.

Proteins can be identified by their characteristic color when they react with specific reagents. In this exercise, we shall attempt to demonstrate some of these reactions.

(A) TO SHOW THAT NITROGEN IS PRESENT IN PROTEINS

Materials Needed: Egg albumin solution, 10% sodium hydroxide, red litmus paper, source of heat, graduate cylinder, test tube.

Procedure:

Place 2 ml of egg albumin solution into a test tube and add an equal amount of 10% sodium hydroxide (made up by mixing 1 part of concentrated NaOH with 3 parts of water). Make certain that the inner rim of the tube is dry and bend a piece of red litmus paper into a U shape. Place the bottom of the U into the mouth of the tube. Then heat the tube gently. When ammonia is freed from the protein solution, the litmus paper will turn blue as the protein breaks down.

(B) TO SHOW THAT SULFUR IS PRESENT IN PROTEINS

Materials Needed: Egg albumin solution, test tube, concentrated sulfuric acid, lead-acetate paper, source of heat, graduate cylinder.

Procedure:

Place 5 ml of the egg albumin solution into a test tube, and add a few drops of concentrated H_2SO_4. Moisten a piece of lead-acetate paper and crumple it into the mouth of the test tube. When the tube is heated gently, the paper will turn black showing the presence of sulfur which is being released.

(C) PROTEIN COLOR IDENTIFICATION TESTS

Materials Needed: Egg albumin, test tubes, test tube rack, concentrated nitric acid, ammonium hydroxide, Millon's reagent, biuret reagent, pepsin, pancreatin, water bath, source of heat, graduate cylinders.

Procedure:

1. Xanthoproteic Test: Place 2 ml of egg albumin solution into a test

tube, add 1 ml of concentrated nitric acid and heat gently. It will coagulate and turn yellow. Now cool under water and then add ammonium hydroxide to the tube. The result will be a bright orange-yellow color.

2. Millon's Test: Place 2 ml of egg albumin solution into a test tube and add 2 ml of Millon's reagent. Warm gently to boiling. A precipitate will form which will turn red. This is due to the hydroxyphenyl group in the protein molecule.

3. Biuret Test: Put 2 ml of egg albumin solution into a test tube. Then add several drops of Biuret reagent. A pale violet color will develop. The Biuret test is very sensitive to proteins and polypeptides. With proteins it turns violet and as the proteins break up into polypetides, the color test will be pink or rose color and finally with amino acids there will be no color change.

4. Hydrolysis of Proteins: As protein is broken down into amino acids, it changes first to polypeptides and then into amino acids. These changes can be demonstrated by means of the Biuret test.

Take 7 test tubes and number them from 1 through 7 and add the following to each:

#1—10 ml of egg albumin solution, 10 ml of 0.5% pepsin solution.

#2—10 ml of egg albumin solution, 10 ml of 0.5% pepsin solution and 2 drops of concentrated HCl.

#3—10 ml of egg albumin solution, 10 ml of 0.5% pancreatin solution.

#4—10 ml of egg albumin solution, 10 ml of 0.5% pancreatin solution plus 2 drops of concentrated HCl.

#5—Just 10 ml of egg albumin solution

#6—Just 10 ml of 0.5% pepsin solution

#7—Just 10 ml of 0.5% pancreatin solution.

Place all 7 tubes in a water bath at 40°C. Wait from 30 to 45 minutes and then examine the contents of the tubes by using the Biuret reagent.

#1—light violet—no hydrolysis

#2—pink—there has been some change to polypeptides

#3—light violet—no hydrolysis

#4—pink—there has been a change to polypeptides

#5—violet—no change

#6—colorless as before

#7—colorless as before.

If tubes 2 and 4 are left in the water bath for a longer period of time, there will be a further breakdown to amino acids and the solutions in the test tubes will turn colorless upon testing with Biuret reagent.

CONCLUSIONS AND DISCUSSION:

1. When the test to show the presence of nitrogen is done, why does the ammonia turning the litmus paper blue prove that nitrogen is present?

2. Explain the chemistry of the reaction which demonstrates the presence of sulfur in proteins.

3. What atomic grouping of protein seems to be responsible for the color reaction in the xanthoproteic test?

4. See if you can find out what the compound *Biuret* is made up of and why this is a test for proteins.

5. Explain all the steps gone through by proteins in their hydrolysis.

THIN-LAYER CHROMATOGRAPHY OF AMINO ACIDS*

AIM: To learn how to identify and separate various amino acids out of a mixture in solution.

We have seen how to identify proteins and how the hydrolysis of proteins results in the formation of amino acids which are called the building blocks of proteins. Now through thin-layer chromatography we have a most modern method of separating and identifying various types of amino acids.

MATERIALS NEEDED: Coated microchromatic slides, Coplin jars, leucine, glutamic acid, aspartic acid, n-butanol, glacial acetic acid, water, ninhydrin, spray, asbestos sheet, hot plate, micropipettes, plastic ruler in centimeters, stoppered bottles, balance, graduates.

PROCEDURE:

1. First prepare as many microchromatic slides as needed. The preparation of these slides is explained in Unit I.

2. Prepare amino acid solution by dissolving 1 mg of the amino acid per ml 0.5 M HCl in an alcoholic solution. Add some water to dissolve if necessary. Label leucine #1, glutamic acid #2, and aspartic acid #3.

3. Prepare a mixture of all three amino acids by mixing together equal amounts of each of the amino acids into another bottle and label #4.

4. Prepare the solvent by mixing together 60 ml of n-butanol, 15 ml of glacial acetic acid and 25 ml of water.

5. Number 4 Coplin jars from 1 to 4 and place about 2 ml of the developing solution into each of the chromatojars. Cover the jar for a few minutes so that the vapors fill the container.

6. Spot the slides by using a different amino acid for each slide as well as the mixture. Thus you will have 4 different slides. Fill a pipette with a solution by capillary action. Touch only the tip of the pipette to the silica gel surface without breaking the surface. Arrange 20 drops adjacent to each other to form a line across the slide about 1 cm from the bottom of the slide. Allow the spots to dry thoroughly and then place the leucine slide #1 in the Coplin jar #1 and so on until you have all 4 slides in the proper chromatojars.

7. The solvent will rise, and when it approaches the top of the silica gel

layer, you remove each slide from the jar. All the slides should be kept in for the same period of time so that all conditions will be nearly the same.

8. Allow the slides to air dry completely and then spray with the visualizer made up of 0.3% ninhydrin in n-butanol containing 3% glacial acetic acid.

9. When the visualizer has dried slightly, place the slides on an asbestos pad on a hot plate which is set at 200°.

10. Within a few minutes the slides will show color. Slide #1, leucine, will show a band that has traveled the farthest. Slide #2, glutamic acid, will show a band that has traveled up about halfway, while slide #3 will show aspartic acid to have traveled the least. As for slide #4 which contains all three amino acids, you will see three bands of various shades of red which will be in the same place as on the individual slides.

CONCLUSION:

Thus we have a method of identifying amino acids.

NUCLEIC ACIDS*

AIM: To test for nucleic acids.

In 1871 Miescher first isolated the cellular substances known as nucleic acids. However, it was not until almost 75 years later that the true role and significance of these nucleic acids were realized.

It is now known that the nucleic acid molecules are the basis for all life processes performed by the cell. It was once believed that DNA (deoxyribonucleic acid) and RNA (ribonucleic acid) were both confined to the nucleus of the cell. However, it is now known that practically all the DNA is confined to the nucleus and only 10% of the RNA is found in the nucleus. The balance of the RNA is found in the cytoplasm, surrounding the nucleus, and comprises the small structures known as ribosomes.

The DNA of the chromosomes is ultimately responsible for the sequence of the amino acids in protein molecules. It is self-replicating and reproduces an exact copy of itself.

RNA is directly responsible for the synthesis of protein molecules and dictates the sequence of the amino acid chain as the protein forms. There are several types of RNA found in the cell and each one is involved in a different role in the synthesis of protein.

We know more about DNA than we do about RNA at the present time. We can identify the two substances within a cell by means of a test called Dische or Diché test. You will use this test in this exercise.

MATERIALS NEEDED: Glacial acetic acid, concentrated sulfuric acid, diphenylamine, distilled water, yeast, 2 test tubes, water bath, test tube rack, 2 graduated cylinders.

PROCEDURE:

1. Make up the Dische reagent by mixing 1.00 gram diphenylamine with 100 ml glacial acetic acid and 2.75 ml concentrated sulfuric acid. Mix to dissolve. Stir in dark bottle at 2°C. Warm to room temperature before use.
2. Place 3 ml of yeast suspension in a test tube and then add 10 ml of the Dische reagent.
3. Heat in water bath for 30 minutes almost at boiling. A marble placed in the mouth of the tube will act as a condenser.
4. Remove from water bath and let stand in test tube rack. There will be a color change. This reagent causes DNA to turn blue and RNA to turn green.
5. You should get a blue color as there is more DNA in the yeast cell than RNA.

CONCLUSION AND DISCUSSION:

1. Why would there be more DNA in a yeast cell?
2. What type of cell would probably show more RNA?
3. Why should the cell you just named contain RNA?

UNIT III

MAINTENANCE IN ANIMALS

The maintenance of animal life is dependent on universal requirements:

- obtaining and distributing essential organic and inorganic compounds.
- removal of metabolic waste products.
- regulation of processes.

I. NUTRITION

A. *Process*: Nutrition includes those activities of animals by which they obtain and utilize various substances (known as nutrients or foods) for metabolic activity.

Foods or nutrients are those substances that serve as:

- energy sources for various life activities of cells
- constituents in the building or repair of cell structures
- regulators of metabolic processes.

Animals lack the ability to synthesize nutrients from inorganic raw materials and must consume preformed organic compounds. This mode of nutrition is known as heterotrophic nutrition.

Usually, other organisms (plants and/or animals) are consumed by animals and serve as a source of preformed nutrients. Food organisms may contain unnecessary compounds as well as those with nutritional value.

1. Ingestion: Ingestion describes those activities of animals by which they take in materials (compounds and/or organisms) which serve as a basis for nutrition.

2. Digestion: Digestion involves the reduction of large or insoluble food materials to small, soluble molecules which may be distributed throughout an animal and used for various physiological activities.

Digestion may occur within cells (intracellular) or outside of cells (extracellular).

a. *Mechanical Aspects*: Many animals exhibit adaptations which permit the breakdown of food materials to small particles through purely physical means such as grinding, cutting, and tearing.

b. *Chemical Aspects*: Mechanical digestion reduces food materials to particles which, by molecular standards, are still exceedingly large. Further reduction to soluble form is accomplished through the action of enzymes.

(1) Enzyme Action: Enzymes are proteins which act as catalysts and affect many physiological reactions, including digestion.

Digestive enzymes are hydrolytic. They promote the breakdown of large molecules to small unit molecules with the addition of water.

The name of an enzyme is usually formed by adding the ending *-ase* to a stem which is taken from the substrate (compound, or class of compound), upon which the enzyme acts. Some examples of digestive enzymes and their substrates include:

Enzyme	*Substrate*
maltase	maltose
lipase	lipids
protease	protein

Enzymes are specific in their action. They will usually only catalyze reactions involving a single substrate compound or a group of closely related compounds.

Temperature and hydrogen-ion concentration (pH) are two of the factors which determine the rate of enzyme action.

(2) Outcome: As a result of the action of enzymes in chemical digestion, complex organic molecules are hydrolyzed to smaller, soluble molecules.

- carbohydrates (sugars, starches, etc.) are converted to simple sugars such as glucose.
- proteins are converted to amino acids
- lipids (fats, oils, etc.) are converted to fatty acids and glycerol.

B. *Adaptation*: Nutritional patterns among animals vary in accordance with the adaptations which have evolved to permit the ingestion and digestion of food materials.

1. *Protozoa*: Protozoa ingest food organisms through fixed openings (as in Paramecium) or by engulfing them (as in Ameba).

Protozoa have no digestive tract. Ingested food materials accumulate in food vacuoles where digestion occurs. Digestion is intracellular.

2. *Hydra*: Hydra possess a digestive cavity with a single opening. The opening functions as a mouth for the ingestion of food and also as an anus for the expulsion of undigested and indigestible materials.

Digestion begins in the cavity (extracellular digestion) where enzymes, secreted from specialized cells of the inner lining, partially hydrolyze the ingested food.

Many of the remaining fragments are engulfed and digested by other cells of the inner lining (intracellular digestion).

3. *Earthworm*: The digestive system consists of a tube within the body cavity with a mouth opening at one end and an anal opening at the other end.

This one-way digestive tube permits specialized parts of the tube to perform sequential digestive functions.

4. *Grasshopper*: In the grasshopper the structure and function of the digestive system are essentially similar to those in the earthworm.

5. *Man*: In man the structure and function of the digestive system are basically like those of the earthworm and grasshopper.

II. TRANSPORT

A. *Process*

1. Absorption: The end products of digestion, as well as other dissolved solids and gases, enter the fluids of animals through membranes.

a. Structure of the Cell Membrane: The cell membrane selectively regulates the entry and exit of materials. This controlling influence of cell membranes aids cells in maintaining homeostasis.

The cell membrane is a complex, porous structure composed mainly of protein and lipid materials.

Dissolved solids (e.g., glucose), gases, ions, and water molecules diffuse through the cell membrane. Large molecules such as proteins, fats, and starch cannot diffuse into or out of cells.

b. Function of the cell membrane: The cell membrane has both passive and active roles in transporting materials into and out of cells.

(1) Passive transport: Some materials enter and leave cells by diffusing through the cell membrane. Diffusion is a form of transport in which cells do not contribute energy for the movement of materials.

Dissolved solids (e.g., glucose), gases, ions, and water molecules move from a region where they occur in higher concentration to a region of lower concentration of the molecules or ions. The diffusion of water into and out of cells is called osmosis.

(2) Active Transport and Pinocytosis: Active transport and pinocytosis are types of transport which require the expenditure of cellular energy.

Active transport is the process in which cellular energy is used to move molecules across membranes against a concentration gradient (i.e., from a region of lower concentration to a region of higher concentration).

Pinocytosis is a process in which large molecules or molecular aggregates that are too large to pass through the cell membrane may be engulfed and brought within the cell. Pinocytosis also requires the expenditure of cellular energy.

2. Circulation: Circulation is the transport of materials within cells and/or throughout multicellular organisms.

B. *Adaptation*: Animals have a variety of adaptations from the absorption and circulation of materials.

1. Protozoa: Protozoa live in water, a transport medium. Absorption occurs between the environment and the cell contents.

Circulation within cells is accomplished by diffusion and by cytoplasmic streaming (cyclosis).

2. Hydra: Since most of its cells are in direct contact with a watery environment, the Hydra has no special transport system.

The processes involved in transport in the Hydra are similar to those of protozoa—diffusion, osmosis, cyclosis, and active transport.

3. Earthworm: Many cells of the earthworm are not in direct contact with the external environment. An internal, closed circulatory system has evolved which indirectly brings materials from the external environment into contact with all cells. This system transports respiratory gases, water, and dissolved solids, including the end products of digestion. An infolding of the earthworm's digestive tube is an adaptation which increases the absorptive surface through which digestive end products enter the blood.

4. Grasshopper: The grasshopper has an internal, open circulatory system which indirectly brings materials from the external environment into contact with its cells. Like the earthworm, the grasshopper has an infolded digestive tube to increase absorptive area.

5. Man: The close circulatory system of man is similar in basic structure and function to that of the earthworm. Man also has an adaptation for increasing absorptive area in the digestive tube.

III. RESPIRATION

A. Process:

1. Cellular level: Cellular respiration (also called internal respiration) occurs continuously in the cells of all animals. Most animals make energy available for cell activity by converting the potential energy of organic molecules (such as glucose) to the more available form of the high-energy compound adenosine triphosate (ATP). ATP synthesis may employ free oxygen (aerobic respiration) or it may not (anaerobic respiration).

a. Aerobic respiration: Most of the enzymes involved in cellular respiration are located in the mitochondria.

(1) Description: During aerobic respiration, the energy of a compound such as glucose is released in small increments by the action of enzymes in removing hydrogen from the molecule. (Removal of hydrogen from molecules is known as oxidation.) The energy released is used to synthesize ATP.

The released hydrogen is combined with oxygen to form water.

(Oxygen serves as a hydrogen acceptor.) The remnants of the original glucose molecule are reorganized to form carbon dioxide.

The overall summary equation for aerobic respiration of glucose is:

$$\text{glucose} + \text{oxygen} \xrightarrow[\text{enzymes}]{\text{numerous}} \text{water} + \text{carbon dioxide} + 38\ \text{ATP's}$$

The energy of the ATP molecules is available for cellular activity and is released through the action of the enzyme ATP-ase.

(2) Chemical aspects: Investigation of the chemistry of glucose oxidation has shown that the process, although involving many enzyme-catalyzed steps, occurs in two basic stages—the aerobic phase and the anaerobic phase.

During the anaerobic phase, which does not involve molecular oxygen, glucose is converted to two molecules of pyruvic acid, and the released energy is used to synthesize 4 ATP molecules. However, since the energy of 2 ATP's is necessary to activate the reactions, there is a net gain of only 2 ATP molecules during the anaerobic phase.

The reactions of the aerobic phase continue the process of energy release by breaking the pyruvic acid down, eventually forming carbon dioxide. During this process, 36 ATP molecules are synthesized.

The net gain from the aerobic respiration of glucose is, therefore, 38 molecules of ATP.

b. Anaerobic respiration: Almost all animals carry on aerobic respiration. However, certain animal cells respire anaerobically when oxygen is deficient or absent.

(1) Description: When it occurs in animals, anaerobic respiration leads to the formation of lactic acid and a relatively small amount of ATP.

Lactic acid production may be summarized as follows:

$$\text{glucose} \xrightarrow{\text{enzymes}} 2\ \text{lactic acid} + 2\ \text{ATP}$$

Lactic acid still contains much unreleased potential energy and the net gain of ATP is 2 molecules. This is a considerably less efficient energy-releasing system than the aerobic respiration of glucose.

(2) Chemical aspects: The formation of ATP in lactic acid production results from the same anaerobic reactions which occur in the anaerobic phase of aerobic respiration. There is no energy yield in the conversion of pyruvic acid to lactic acid.

The reactions may be summarized as follows:

$$\text{glucose} \xrightarrow[\searrow 4\ \text{ATP}]{2\ \text{ATP} \searrow} 2\ \text{pyruvic acid} \xrightarrow{\text{enzyme}} 2\ \text{lactic acid}$$

Deducting the 2 ATP's necessary for the activation of the process yields a net gain of 2 ATP's from the reactions which lead to the synthesis of lactic acid.

2. Organism level: Since anaerobic respiration is rare among animals, most respiratory activity requires oxygen and is essentially the same in all animal cells.

The major differences in overall respiratory activity among different animals is in the manner in which they acquire oxygen for respiration and the method used to remove the carbon dioxide resulting from respiratory activity.

B. Adaptations: Animals have a variety of adaptations for the exchange of respiratory gases with their environment.

1. Protozoa: In Protozoa, respiratory gases are exchanged directly with the watery environment by absorption through cell membrane.

2. Hydra: The Hydra, like the Protozoa, is completely surrounded by water. Gas exchange occurs directly with the environment by absorption.

3. Earthworm: In the earthworm, mucus secreted by cells of the external body wall provides a moist surface for gas exchange with the environment. The circulatory system transports respiratory gases to and from the moist skin (external respiration).

4. Grasshopper: The grasshopper has evolved an independent system of respiratory tubules for the intake, distribution, and removal of respiratory gases. The tracheal tubules are greatly branched. Their endings are intimately associated with moist internal tissues where gas exchange takes place.

5. Man: Both man and the grasshopper have evolved a separate system of respiratory tubes. In man these are also greatly branched and extend throughout specialized gas exchange organs called lungs. The lungs also contain arteries, arterioles, veins, venules, and capillaries of the circulatory system.

Gas exchange occurs between the moist respiratory tubules and the capillaries.

IV. EXCRETION

A. *Process*: The metabolic wastes of animals include water, carbon dioxide, mineral salts, and nitrogenous wastes.

The metabolic activities which produce these wastes include:

Metabolic activity	*Waste(s) produced*
Respiration	Carbon dioxide and water
All metabolic processes	Mineral salts
Dehydration synthesis	Water
Protein metabolism	Nitrogenous wastes

B. *Adaptations*: Animals have a variety of adaptations for the disposal of harmful metabolic wastes.

1. Protozoa: In Protozoa, carbon dioxide and mineral salts diffuse directly into the watery environment.

Since Protozoa are surrounded by fresh water, removal of water occurs against a concentration gradient. Active transport, including the "pumping" action of contractile vacuoles, serves in the excretion of excess water.

Nitrogenous waste, in the form of ammonia (which is very soluble in water) diffuses through the cell membrane.

2. Hydra: Excretion in the Hydra is essentially similar to that in Protozoa.

3. Earthworm: Carbon dioxide is excreted by diffusion through the moist skin of the earthworm.

Water, mineral salts, and nitrogenous wastes (in the form of ammonia and urea) are excreted by the action of paired nephridia located in each body segment.

4. Grasshopper: Carbon dioxide diffuses from the grasshopper's blood into tracheae where it is transported to the atmosphere.

Water, mineral salts, and nitrogenous wastes (in the form of uric acid) accumulate in excretory tubules (Malpighian tubules) and are transported to the digestive tube. Most of the water is reabsorbed. Minerals and uric acid are expelled along with fecal material.

5. Man: Carbon dioxide is excreted by the action of the lungs.

The principal nitrogenous waste in man is urea. Urea, water and mineral salts are removed from the blood in the nephrons of the kidneys. A system of tubes transports these wastes out of the body.

V. SYNTHESIS

A. *Process*: All animals, after processing ingested materials, use the resultant simple compounds for the synthesis of the thousands of compounds needed for maintenance.

Synthesis activities are many and varied. However, one common synthetic mechanism is known as dehydration synthesis. In this activity, simple unit molecules, such as glucose or amino acids, are incorporated into complex molecules made up of large numbers of unit molecules. In the process water is released.

Dehydration synthesis accounts for the synthesis of glycogen from glucose and proteins from amino acids. Glycogen, also known as animal starch, resembles the starch which plants produce.

B. *Products*:

1. *Secretion*: Although synthesis occurs in all cells, certain substances, which are synthesized in abundance and have special roles in metabolism, are known as secretions. These are often the result of activity in specialized cells or tissues.

Among the most abundant secretions are enzymes, hormones, and neurohumors.

a. Enzymes: Enzymes are proteins which catalyze virtually all reactions occurring in cells. Since the relationship between enzymes and the reactions they catalyze is highly specific, thousands of different enzymes are synthesized.

b. Hormones: Hormones are compounds which vary considerably in their chemical properties. In general, they regulate a wide variety of physiological processes by accelerating or retarding activity.

c. Neurohumors: Neurohumors function in much the same manner as hormones. However, they are secreted by nerve cells.

d. Other: In addition to the secretions mentioned above, animals produce a wide variety of other specialized secretions. Examples include:

- hydrochloric acid of human gastric glands
- poisons secreted by various animals
- mucus
- human lachrymal fluid (tears)
- waxes and oils

2. *Structural compounds*: All of the structural components of cells result from the organization of synthesized materials into special, fixed relationships. Example: protein and lipid organization in membranes.

C. *Outcome*: The results of synthesis activity are many and varied. Growth and repair of structures are two of the important results.

In addition, the production of food reserves is a result of synthetic activity. Fat and glycogen are the most common storage products in animals.

VI. REGULATION

A. *Process*: Organisms receive information from their environment and may respond to it by a movement, a secretion, or growth.

In most animals regulation is achieved by the interaction of a nervous system and a system of specialized secretions.

1. Nervous control

a. Stimulus and response: A stimulus is a change in the environment that brings about a response. Stimuli may be external or internal.

Responses are the reactions of organisms to stimuli.

In most complex animals a system of specialized cells (neurons) reacts to stimuli from the environment. Stimuli initiate electrochemical impulses which are transmitted to the site of the response by a system of neurons.

b. Transmission of nerve impulses: Nerve impulses are transmitted through a system comprising neurons, synapses, and neurohumors.

(1) Neurons: Neurons are specialized cells which may be greatly elongated, thus providing rapid transmission of impulses over long distances.

(2) Synapses: Synapses are junction spaces between adjacent neurons.

2. Endocrine control: Regulation of many animal functions is accomplished by means of hormones. Glands which secrete hormones discharge their secretions directly into the circulatory fluid and are called endocrine or ductless glands.

a. Hormones: Hormones are sometimes called "chemical messengers." They are secreted by specialized cells and control the activities of other cells.

b. Effects: Hormones affect many activities in animals including:

- growth patterns and rates
- metamorphosis in some animals
- periodic behavior, e.g. mating activity

3. Comparison of Nervous and Endocrine Systems: The nervous and endocrine systems have certain similarities and certain differences.

Two similarities are:

- Both play a major role in maintaining homeostasis.
- Both secrete chemical messengers.

Three differences are:

- Nerve responses are more rapid than endocrine responses.
- Nerve responses are of shorter duration than endocrine responses.
- Nerve impulses travel by way of neurons while hormones are carried by the transport system.

B. *Adaptations*

1. Nervous System: Animals have evolved a multiplicity of adaptations which tend to improve their ability to respond to stimuli.

a. Protozoa: Protozoa exhibit regulated activity in response to environmental stimuli.

b. Hydra: The nervous system of the Hydra comprises a nerve net consisting of modified neurons in which impulses may travel in either direction.

c. Earthworm: The nervous system of the earthworm consists of a relatively simple anterior "brain" (fused ganglia), a ventral nerve cord, and connecting nerves.

The presence of a central nervous system permits impulses to take a definite pathway from receptors to effectors.

d. Grasshopper: The structure and function of the grasshopper's nervous system is similar to the nervous system of the earthworm.

e. Man: The nervous system of man, composed of a brain, dorsal nerve cord and nerves, consists of two major functional divisions:

- The central nervous system is involved with both voluntary and involuntary behavior.
- Many hormones have wide distribution among animals, the same hormones occurring in both simple and highly complex animals.

VII. LOCOMOTION

A. *Process*: The ability to move from place to place has increased the possibility of animal survival. Locomotion offers a number of advantages including:

- increased opportunities to procure food
- ability to seek shelter
- movement away from toxic wastes
- escape from predators
- increased likelihood of contact between individuals capable of mating with each other.

B. *Adaptations*: Animals have a wide variety of locomotive adaptations which permit survival in their particular environments.

1. Protozoa: The locomotive structures utilized by Protozoa include cilia, flagella, and pseudopodia.

2. Hydra: The Hydra is essentially a sessile organism (i.e., it tends to remain in a fixed position much of the time). However, the presence of contractile fibers permits some motion including a type of somersaulting motion by which it changes position at times.

3. Earthworm: Locomotion in the earthworm is accomplished through the action of muscles and bristles (setae). The setae produce temporary anchorage in soil as muscles produce alternating extension and contraction of the animal's body.

4. Grasshopper: Locomotion in the grasshopper is accomplished by interacting muscles which operate chitinous appendages (legs and wings).

5. Man: Locomotion in man is accomplished by the interaction of muscles which move the bones of the skeleton.

DIFFUSION THROUGH A MEMBRANE

AIM: To study diffusion through a nonliving membrane in order to understand diffusion in living cells. Also to study the action of an enzyme and its effect on diffusion.

Diffusion is a form of transport in which cells do not contribute energy for the movement of materials. However, movement is the result of the kinetic, or heat, energy of a system of particles in motion.

Diffusion is the process in which ions and soluble molecules move from a region where they occur in higher concentration to a region of lower concentration of the molecules or ions. The diffusion of water into and out of cells is called osmosis.

MATERIALS NEEDED: Soluble starch solution; 80% glucose solution; iodine solution; distilled water; Benedict's solution; Bunsen burner; 3 pieces of cellophane dialysis tubing, about 6 inches each; test tubes and holder; 1 ml pipette; ring stand; clamp; meter stick; thermometer; graduate; medicine dropper; string; rubber bands; small narrow-mouth jars.

PROCEDURE:

(A) Diffusion of Starch and Glucose:

1. Soak the dialysis tubing thoroughly in water before using. Tie a knot at one end of the tubing and fill the tube to about one inch from the top with the starch solution. Add 20 drops of glucose solution.

2. Tie the top of the tubing with string, rinse under cold water so that there is nothing on the outside and place into a beaker of water to which you previously added about 5 ml of iodine for each 50 ml of water in the beaker.

3. Let it stand for about 25 to 30 minutes. You will then notice that the iodine has diffused into the cellophane bag because there will be a brownish color at the bottom of the bag. Why does this occur?

4. Now take a little of the water in the beaker. Place in a test tube and test for sugar with the Benedict's solution. The material in the test tube will turn brick red upon heating, which shows the presence of sugar.

5. Since the cellophane bag has become swollen, we assume that the water, glucose, and iodine diffused through the membrane while the starch did not. However, test for it with Benedict's solution.

(B) Effect of An Enzyme on Diffusion:

1. Chew on a sterilized rubber band to facilitate the flow of saliva and collect about 10 ml in a test tube.

2. Take a piece of the soaked dialysis membrane and knot at one end. Place the saliva and 10 ml of the starch solution into the membrane and place the filled bag into a small narrow-mouth bottle so that the top of the membrane can hang loosely over the top of the bottle. Fill the bottle with about 50 ml of water or enough to reach over the contents of the bag.

3. In about 25 minutes, take some of the contents of the bottle and put into each of two test tubes. Test one for sugar with the Benedict's solution and one for starch with the iodine solution. There will be a positive test for sugar and a negative for starch showing that when the starch changed to sugar through the use of saliva, the sugar then diffused through the membrane.

(C) Diffusion Pressure

1. After soaking the dialysis membrane, tie a knot in one end and fill the tubing with glucose solution. Place a 1 ml pipette into the tubing and close the top of the tube around the pipette by means of a rubber band.

2. Fasten the top of the pipette into a clamp which is attached to a ring stand while the bottom of the tubing is immersed in a small bottle filled with distilled water. Now place a meter stick along the side of the pipette.

3. Record the initial height of the liquid in the pipette and then take readings every 3 minutes. There will be a rapid rise of liquid in the pipette showing that water diffuses very rapidly into the tubing.

CONCLUSION:

Answer the following questions.

1. In Part A of the exercise, what has happened to the various materials?
2. Which material was not able to diffuse through the membrane? Why?
3. In Part B of the exercise, what materials diffused in opposite directions?
4. Why was sugar present in the water of the bottle in part B?
5. In Part C, why did water diffuse more rapidly into the tubing than the glucose diffused out of the membrane?

SELECTIVE ACTION OF CELL MEMBRANES*

AIM: To study diffusion and the semipermeability of the living cell membrane in an actual cell.

The membrane of a living cell plays an important part in the regulation of the passage of materials into and out of the cell. Only certain substances can enter and leave the cell and thus a chemical balance is maintained with its surroundings.

In this exercise we shall see what happens while the membrane is living and what happens when the organism dies.

MATERIALS NEEDED: Suspension of yeast cells prepared by adding the powdered yeast to a 10% solution of molasses and allowing it to incubate at 30°C for about 24 hours before use; Congo red solution, microscope, slide and cover glass, test tubes and test tube rack, Bunsen burner, test tube holder, and pipette or medicine dropper.

PROCEDURE:

1. Into each of the two test tubes place approximately 1 ml of the yeast suspension.
2. Add 3 drops of the Congo red solution to each test tube.
3. Take a drop of the suspension and study under the compound microscope. If you use the high-power you will notice that the cells are surrounded by the Congo red. None of the dye entered the yeast cells.
4. Now heat the second tube over a flame until it reaches boiling point. Cool and study a drop of this suspension under the high-power magnification. You will notice that there is a red color throughout and inside the cells which means that the cell membrane no longer was there to prohibit the passage of the dye.

CONCLUSION:

1. What happened to the membrane surrounding the yeast cell when the yeast was boiled?

* Starred experiments throughout the book should be checked for materials at least two days in advance. Some materials must be prepared ahead of time.

2. Does a cell have to be alive for its membrane to function?
3. What would you say about the activity of cell membranes?
4. What would you suppose are the relative sizes of the molecules of Congo red and the pores of the cell membrane?

PROTOZOA

AIM: To study the simplest animals found and to observe their activities.

Protozoa have single cells and yet they are highly successful at staying alive in their watery environment. Each cell or animal carries on the same metabolic functions as do the many cells of larger animals.

MATERIALS NEEDED: Microscope, slides, cover slips, pipettes or medicine droppers, cotton fibers, methyl cellulose, ameba, euglena, paramecium cultures.

PROCEDURE:

(A) Ameba

1. Place a drop of sediment from the ameba culture on a slide and place a cover glass over it. Examine under low power and observe the movement of the organism. You will notice that lobes of cytoplasm move forward and propel the animal. These lobes of cytoplasm are called pseudopodia and more than one can be found on a single animal at a time.
2. If you watch long enough you will see the pseudopodia of the ameba surround and ingest a particle of food.
3. You will notice the contractile vacuole which is used to remove the wastes formed in the animal. You will also see a nucleus.
4. Perhaps if you watch long enough, you may find an ameba that seems to be pinched in two with a thin connection between the two parts. This ameba is reproducing by means of fission.
5. In order to study the ameba in greater detail, turn to high power. Study the physical nature of the ectoplasm and the endoplasm. You will see that the ectoplasm is clear and watery while the endoplasm is dense and has tiny spots throughout.
6. Make a drawing of the ameba and label the plasma membrane, pseudopodium, endoplasm, ectoplasm, food vacuole, contractile vacuole, nucleus.

(B) Paramecium

1. Place a drop of the scum from the paramecium culture on a slide. Add several threads from a piece of tissue. Examine under low power and observe the organisms as they swim rapidly about the slide. Why is it called the "slipper animal"? Do they have front and back ends?

2. Watch the manner in which they swim. They rotate as they swim and they swim in a spiral path. When they reach an obstacle, they either attempt to wiggle through or else back off and turn about to swim in a new direction.

3. Now add a drop of methyl cellulose to the slide to slow them down further. Replace the cover glass and examine under high power. How does the paramecium propel itself through the water?

4. Locate the orgal groove and watch for a while until the paramecium ingests a piece of food from the scum. How does this take place?

5. How do the food vacuoles move within the cell? You will find the contractile vacuoles at either end of the cells with the canals radiating outward. Can you find the anal pore? Observation will show you where the material is discharged from the organism.

6. Can you locate the trichocysts just inside the pellicle of the cell? What is the function of the trichocysts?

7. Make a drawing of the paramecium, labeling all parts.

(C) Euglena: This organism possesses both plant and animal characteristics. It is green in color since it can manufacture its own food.

1. Place a drop of Euglena culture on a slide. Cover with a cover glass and examine under low power of the microscope.

2. You will notice that the organism moves about by extension and contraction. This movement is called euglenoid movement. The cell swims about by means of a flagellum as it moves forward.

3. Study the slide under high power and you will notice that there are green structures within the cell. What are they called and what is their function?

4. You will find an eyespot which is necessary for the organism to orient to the light.

5. There is a gullet at the anterior end. Is it used by the organism?

6. Make a drawing of the organism and label anterior end, posterior end, flagellum, gullet, eyespot, chloroplast, and nucleus.

CONCLUSIONS:

Answer the following questions:

1. What structures are involved in carrying out each of the following activities in all three of the organisms you observed? food getting, digestion, circulation, excretion, locomotion, sensation, reproduction, water balance.

2. Is there any evidence that structure and function are related in each of the three organisms?

INTRACELLULAR DIGESTION IN PARAMECIUM*

AIM: To observe ingestion and digestion in Paramecium.

INTRODUCTION: Not being able to synthesize their organic compounds from simple inorganic materials, heterotrophs take in preformed organic compounds. These may be digested outside the cell, as in bacteria and yeasts, or may be digested intracellularly as in Paramecium. In this exercise you will observe how a paramecium ingests yeast cells and you will note changes inside the food vacuoles during the process of digestion.

PREPARATION FOR LABORATORY: The yeast cells which will be fed to the Paramecium have been prepared by boiling them in water to kill the cells so that their membranes will be permeable to the indicator Congo Red. Congo Red is an indicator that is red at pH 5 and blue at pH 3.

PROCEDURE:

(1) Make a ring of methyl cellulose, about ¾″ in diameter, in the center of a microscope slide. Place a drop of Paramecium culture inside the ring and add a cover slip. Focus with care in the usual manner with low and then high power on your microscope. The syrupy methyl cellulose will soon diffuse into the water and slow down the organisms. Observe their general structure and movement. It will be similar to the observation you made previously on Paramecium.

(2) Now remove the cover slip and add a drop of suspension containing the dyed yeast cells. Cover and observe immediately. You should note the food vacuoles containing colored red yeast cells as the food is being swept into the body of the organism by its cilia. Follow the path of a food vacuole for approximately 5 minutes while looking for changes in the color of the yeast cells. As the food is being digested the yeast cells will slowly change from red to blue and you will be able to follow the movement of the yeast. You will note that the movement follows the regular movement of the protoplasm of the organism. Watch and see whether anything is being formed at the oral groove. If you watch long enough, you might notice material being egested from the Paramecium.

OBSERVATIONS AND QUESTIONS:

1. On Diagram 16 use circles to represent food vacuoles and indicate by arrows the path of the food vacuole.

2. What name is given to organisms which can prepare their organic compounds from simple inorganic substances?

3. Organisms which cannot synthesize organic compounds from simple inorganic substances are called . . . . . .

4. Is Paramecium an autotroph or a heterotroph? . . . . . .

5. Define digestion . . . . . .

Diagram 16.

6. Where does digestion occur in bacteria and yeasts? . . . . . .

7. From the chemical viewpoint, digestion consists of the (hydrolysis) (dehydration synthesis) . . . . . . of larger molecules.

8. What structures are used by Paramecium for locomotion? . . . . . .

9. Two structures used in food getting are . . . . . .

10. In what structures does digestion occur in Paramecium? . . . . . .

11. Is digestion in Paramecium intracellular or extracellular? . . . . . .

12. What was the color of the yeast cells immediately after ingestion? . . . . . .

13. What was the first color change to occur? . . . . . .

14. What change in acidity within the food vacuole does this indicate? . . . . . .

15. What additional color changes occurred? What does this indicate? . . . . . .

16. Two sets of enzymes function in Paramecium. The first operates in a(n) . . . . . . medium, the second in a(n) . . . . . . medium.

17. Where are the enzymes produced? . . . . . .

18. Food consisting of bacteria and other microorganisms is swept into the . . . . . . by means of cilia, and a small . . . . . . forms at the base of the gullet. The vacuole is then set free in the cytoplasm. In the vacuole digestion proceeds first in a(n) . . . . . . medium and then in a(n) . . . . . . medium.

The digested food is absorbed into the cytoplasm during cyclosis of the food vacuole, and the undigested food residue is discarded to the outside at the . . . . . . The organic molecules which were formerly part of the bacteria are changed into Paramecium protoplasm under the influence of the hereditary material of the Paramecium, its . . . . . .

This laboratory exercise is used in Far Rockaway High School, Far Rockaway, New York.

HYDRA

AIM: To study the anatomy and behavior of Hydra.

INTRODUCTION: Hydra is a coelenterate that may be found in ponds and slowly moving streams attached to vegetation by means of the basal disc. The body is composed of two layers, the *ectoderm* and the *endoderm.* It can contract or expand the body in response to stimuli. Its nervous system consists of a nerve net in the layer between the cell layers. When living organisms strike the tentacles, the nematocyst cells evert a stinging thread which can paralyze small prey. Glutathione, a substance released from the tissues of the prey, acts as a stimulus to initiate the feeding reflex whereby the tentacles contract, bearing the prey to the mouth which expands to engulf the organism. Digestion occurs within the gastrovascular cavity. Undigested food is egested through the mouth.

Hydra may be fed in captivity upon brine shrimp larvae which are raised daily from brine shrimp eggs. Well-fed Hydra bud profusely. A high concentration of CO_2 gas in the culture medium acts as a stimulus to initiate sexual reproduction.

PROCEDURE:

PART I: ANATOMY

Observe a prepared microscope slide under the low and high power of the microscope. Study both the longitudinal and cross sections carefully and draw several different areas to show the types of cells.

PART II: BEHAVIOR

(1) Use a dissecting needle to touch various portions of a living Hydra which has been placed in a well on a slide made with silicone gum. Record your observations.

(2) Use a medicine dropper to feed the Hydra in the well. Take up a brine shrimp larva and place it next to the Hydra and view what happens under the low power of your microscope or under a binocular dissecting microscope. Describe what you see.

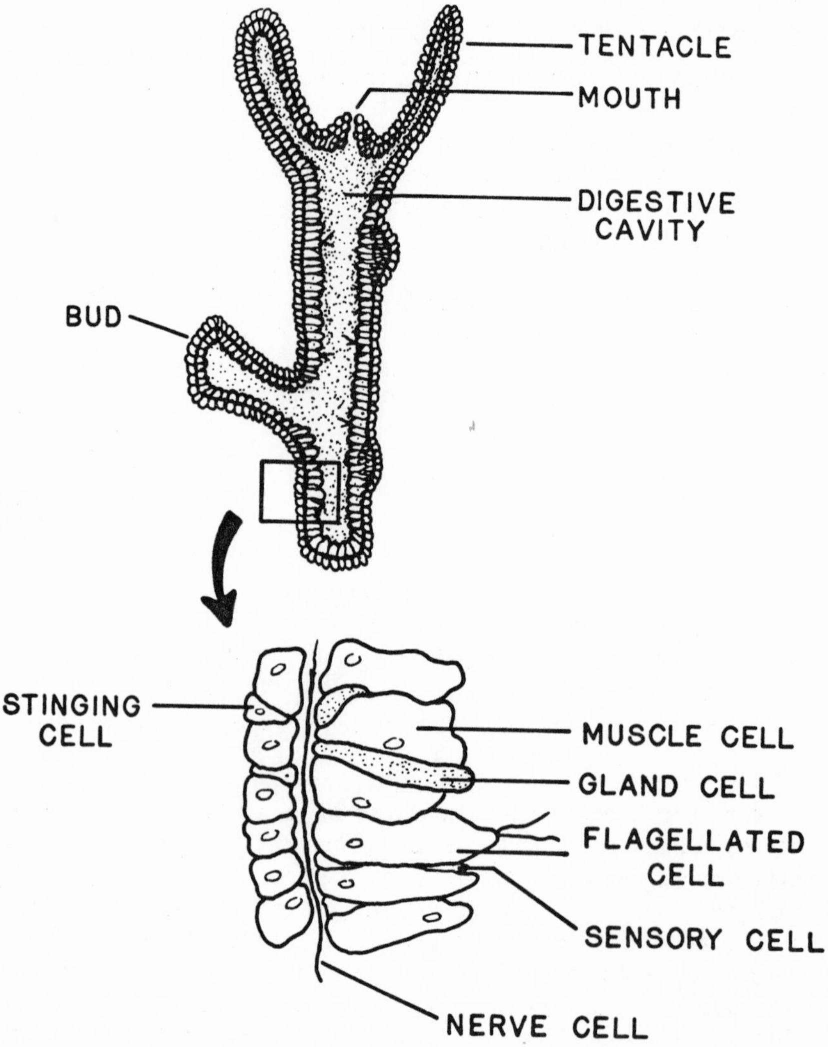
TENTACLE
MOUTH
DIGESTIVE CAVITY
BUD
STINGING CELL
MUSCLE CELL
GLAND CELL
FLAGELLATED CELL
SENSORY CELL
NERVE CELL

Diagram 17.

(3) In another well made of silicone gum place a few drops of dilute acetic acid and then add a Hydra. Observe the results under the microscope. The nematocysts should explode.

OBSERVATIONS AND CONCLUSIONS:

1. Make up a report of what you saw. Include drawings. (See Diagram 17.)
2. Answer the following questions:
   a. To what phylum does Hydra belong?
   b. Name two relatives of Hydra.
   c. How big is this organism?
   d. Where does it live?
   e. Describe its digestive system.
   f. How does it capture food?
   g. How many layers of cells does it have? Name them.
   h. How does it get rid of undigested food?
   i. How does it reproduce?

THE DAPHNIA*

AIM: 1. To learn the structure of a complex invertebrate animal.
2. To study the beating of the heart in a live animal.
3. To study the effects of various chemicals on the heartbeat.

INTRODUCTION:

The invertebrate animals are those without backbones. They make up the major number of animals on earth. They range in complexity from one-celled animals to the enormously diverse and abundant insects and the highly developed molluscs, such as oysters, clams, and the octopus. Daphnia belongs to the group of Crustacea, which, as the name suggests, have a hard shell covering their body. Most crustaceans have five pairs of jointed legs. The Daphnia in this respect differs from its relatives. When you examine the animal you will discover the difference. Most crustaceans live in water. Daphnia is a small crustacean found in fresh water ponds. Although it is called a water flea, it is not a flea. It is an important food for small fish.

Because Daphnia is small enough to be transparent, it can be studied alive under the low power of the microscope or under a dissecting microscope.

MATERIALS NEEDED: One silicone gum living cell with a daphnia in it; carmine suspension, 10% methyl cellulose, dropping bottle of 1% urethane, dropping bottle of 0.01% adrenalin.

PROCEDURE:

Place the living cell containing the Daphnia under low power of a compound microscope or under your dissecting microscope. It would be advisable to add one drop of methyl cellulose to the slide so as to slow down the animal.

1. Answer the following questions as you study the organism:

a. How large is it? Under low power, you can calculate the actual size of the animal. How does it compare with the Paramecium in size?

b. In the head you will see a black area. This is the eyespot. It is not a true eye, but it is sensitive to light. If you were to change the brightness of the light you would notice that the eye rotates and that variation in the intensity of the light changes both the rate and direction of the rotation. Does the eye respond faster to high or low intensity?

c. Find the long *antennae* and shorter *antennules.* If you study them you should be able to determine their function. The antennae are used for swimming, while the shorter antennules are sensory. Do you know of any other organisms that have similar organs?

d. The Daphnia eats continually. It can be fed by adding a drop of carmine suspension to the slide. Watch the movement of the food through the intestine. The food passes through in a movement similar to that in man. What is this movement called?

e. If you watch for several minutes you will see excretion take place through the ventral anus.

f. There are 5 pairs of legs located under the shell that beat continuously. Using the carmine suspension you will notice that the resulting water currents bring oxygen to the gills and food particles to the mouth.

2. Just back of the eyespot and beneath the back of the shell is the pulsating heart. In order to study the heart properly, it would be best to transfer the Daphnia to a clean silicone cell or else start with a fresh organism in the silicone cell. Count the number of beats per minute. It usually pulsates between 180 and 350 times a minute, depending upon the temperature.

3. From the dropping bottle add a drop of 1% urethane to the cell. Reexamine the organism for the effect on the heartbeat. It will have slowed down since urethane is a depressant. In fact, it will slow down the body movements similarly to methyl cellulose.

4. Now add a drop of adrenalin to the cell and reexamine the rate of the heartbeat. You will find that the heartbeat has quickened, because adrenalin is a stimulant.

5. There might be black spherical objects in the hind end of the body. These are eggs.

The "sand fleas" that are found at the beach during the summer are relatives of the Daphnia. These animals have five true legs, by means of which they can dig into the sand with astonishing speed. They are dull-gray in color and about one-half to one inch long.

OBSERVATIONS:

Label the part to which each arrow points.

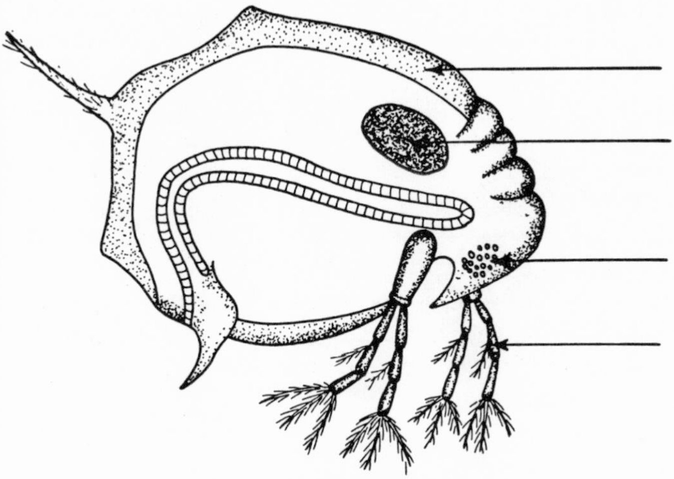

Diagram 18.

THE EARTHWORM

AIM: To study the internal and external features of a segmented worm such as the earthworm.

INTRODUCTION: The earthworm is an annelid whose most noticeable characteristic is the ringed body. Each ring is called a segment. Many of the inner parts of the body are arranged according to the segments. The earthworm lives mostly in soil, having hairs, or setae, which help in crawling. Earthworms are very important in improving the soil for the growth of plants because they help break up the dead matter. They also help make the soil more porous and bring air into the soil to be absorbed by roots. Earthworms bring up the lower soil and deposit it on the surface.

PROCEDURE:

I. EXTERNAL ANATOMY: Since preserved specimens are used, make certain that the specimen is well washed under running cold water to get rid of all traces of the formaldehyde.

1. Note the segmentation in this "segmented worm." When you open your specimen, you will see that this segmentation is internal as well as external.

2. Distinguish between the head end, with its overhanging flap of tissue, and the anus which is merely a slit.

3. The appendages in the earthworm are reduced to hair-like bristles called *setae*. There are four pairs, as indicated in the diagram of the cross section (Diagram 19).

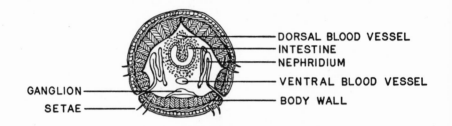

Diagram 19.

4. In segments 31–36 there is present the *clitellum* as a raised portion of the outer wall. This produces a capsule which proceeds anteriorly to gather up the eggs and sperm into a cocoon.

5. The female pores are on segment 14. These are the openings of the oviducts. The male pores on segment 15 are the openings of the sperm ducts. On segments 9 and 10 are the openings through which sperm enter the worm enroute to the seminal vesicles.

II. INTERNAL ANATOMY: Pin the specimen in the dissecting pan with the dorsal side up and with a pin in segment 1 or 2, to the right of the midline. Using a sharp scalpel or scissors, make an incision in the dorsal side, about an inch posterior to the clitellum, slightly to the side of the midline. Extend the cut to the mouth. After you have cut a few inches, check to ensure that you have not gone so deep as to enter the intestine.

Notice the *septa*, or partition, between the segments. Needles may be used to cut the septae. Pin the sides of the wall down as you proceed, placing the pin at an angle with the head end out. Keep the specimen moist with water.

1. *Digestive system*: Note the buccal cavity, pharynx (to segment 15), esophagus, crop (thin-walled, dark in color), gizzard (white and thick-walled), intestine and anus. Note the "tube-within-a-tube" pattern and the

specialized regions, as opposed to the generalized sac-type structure of Hydra.

2. *Reproductive system*: The large masses in segments 10, 11, 12 are the seminal receptacles which receive sperm from the partner during copulation.

3. *Circulatory system*: The dorsal blood vessel may be dark brown in color and it may appear along the middle of the intestine. Look for the aortic arches in segments 11 to 6.

4. *Excretion*: The nephridia are delicate tubes with ciliated openings which are present in pairs in most segments. Observe the demonstration of nephridia under the binocular microscope.

5. *Nervous system*: Remove a section of intestine and look for the solid ventral nerve cord as a delicate white thread. Observe a demonstration of the supraphayngeal ganglion, or brain.

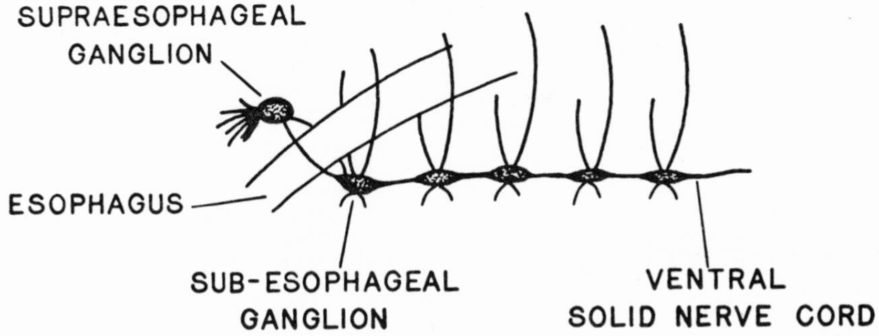

Diagram 20.

6. *Respiration*: Lumbricus obtains oxygen and gets rid of carbon dioxide by diffusion through the moist outer membrane. Capillaries lying below the skin are part of the closed circulatory system. The oxygen combines with hemoglobin.

QUESTIONS:

1. To what phylum does the earthworm belong?
2. What is its genus?
3. Do the external markings correspond with the internal segmentation of this annelid?
4. Why does the earthworm "need" a transport system while Hydra can depend upon diffusion to pass food and oxygen to all cells?
5. How is it possible for the grasshopper to get along with an open circulatory system and with blood lacking hemoglobin whereas the lowly

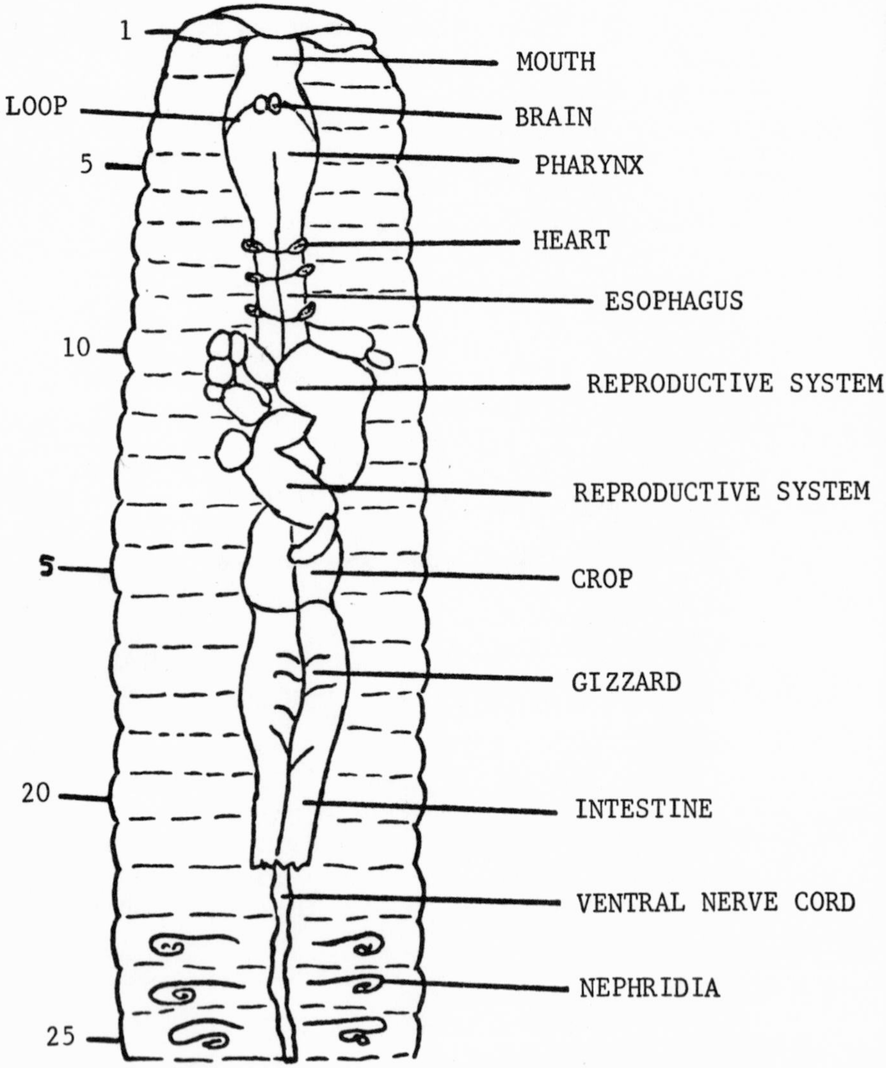

Diagram 21. Internal Anatomy of the Earthworm.

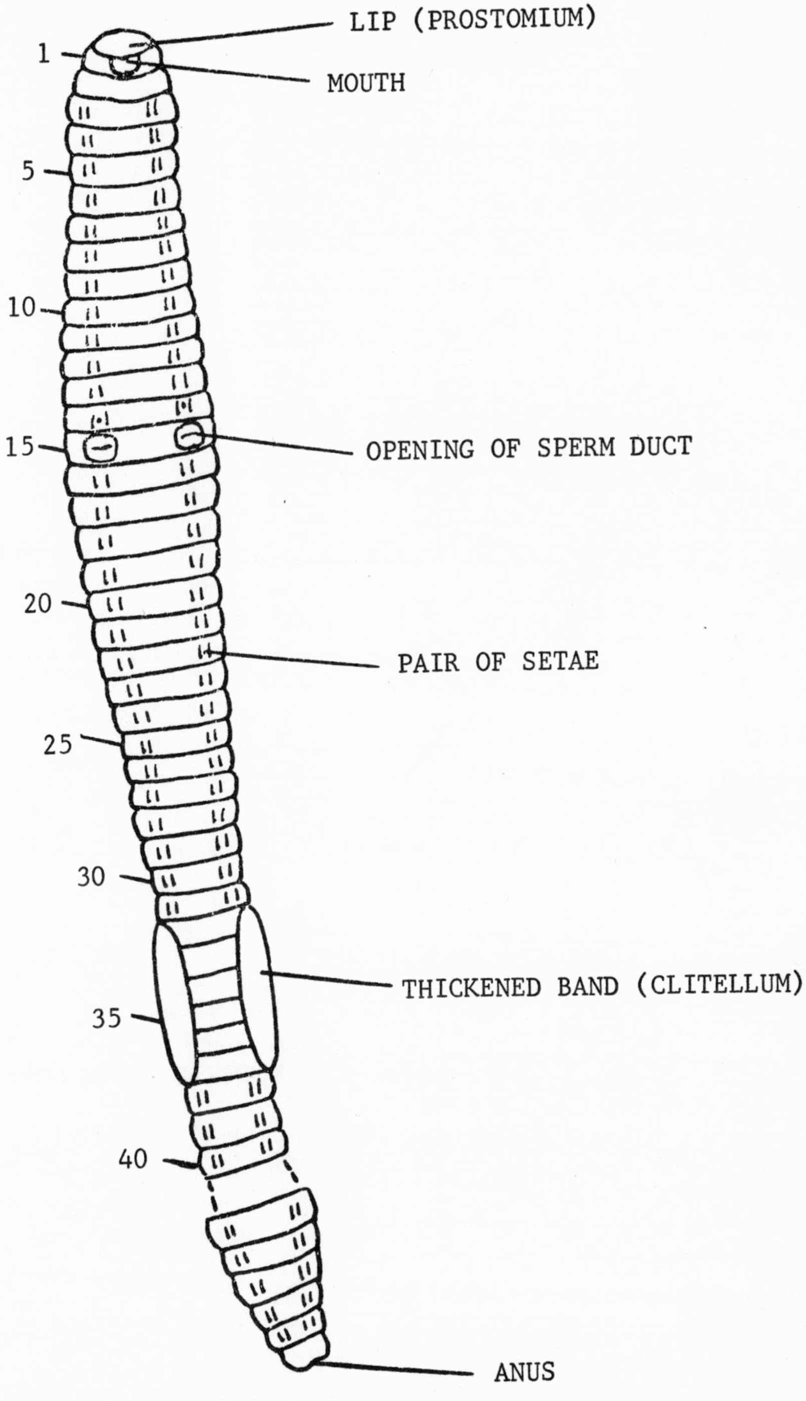

Diagram 22. Ventral View of the Earthworm.

earthworm has a closed system and hemoglobin?

6. What pumps blood in the earthworm?

7. In the earthworm the nerve cord is solid; in vertebrates it is . . . . . . In the earthworm the nerve cord is ventral; in vertebrates it is . . . . . .

8. What structures in the earthworm carry on the function of the vertebrate kidney?

GRASSHOPPER

AIM: To study the external and internal structure of a grasshopper which is classified as an insect.

There are about 5 times as many species in the class Insecta as in all the other animal groups combined. The insects, as a group, are the most completely adapted to life on land.

Insects have an external skeleton and jointed legs as do other arthropods in which group the insects are classified. They have three pairs of legs attached to the thorax. The thorax is the middle of the 3 body regions, the other two being the head and the abdomen. Usually there are 2 pairs of wings attached to the thorax. A pair of antennae and a large pair of compound eyes are found on the head. The respiratory system consists of paired openings leading into a branching system of tracheal tubes.

MATERIALS NEEDED: Preserved grasshopper, waxed dissecting pan, scissors, forceps, microscope slides and cover slips, dissecting microscope, medicine dropper, hand lens.

PROCEDURE:

(A) External study:

1. Use a hand lens to examine the head of the grasshopper. Locate the antennae, the compound eyes and the three simple eyes.

2. Notice where the wings are attached. Do the wings differ from one another? Why would there be a difference?

3. Examine the legs and notice that each leg has hooks on the last segment for holding on to the leaves of plants.

4. Carefully examine the mouth parts. Locate the upper lip, labrum; the lower lip, labium; the maxillae or the little jaws; and the mandibles, large toothlike jaws. Seize the base of the labrum with your forceps and remove it by pulling sharply upward. Then remove the labium by seizing it firmly and pulling sharply downward. Remove the mandible and maxillae. Examine all the parts under the dissecting microscope.

5. Now examine the abdominal segments by removing the wings. Locate the tympanic membrane, or ear, on the first segment. Find the spiracles,

openings into the respiratory system, which are on the side of each abdominal segment. At the extreme tip of the abdomen there will be a very hard, pointed ovipositor if your specimen is female.

(B) Internal Study:

1. Place the grasshopper in the dissecting pan. Make an incision along one side of the dorsal midline. Begin at the extreme posterior end and work forward into the thorax. Do not insert the tip of the scissors too deeply. As you cut anteriorly, pin the lower edge of the incision to the dissecting pan wax. Make a similar incision on the opposite side and remove the entire dorsal body wall.

2. If your specimen is female it will contain masses of large yellow eggs. Remove the eggs. If your specimen is male the testes are located in the posterior portion of the abdomen. The testes are white or yellow in color.

3. Notice the elongated heart, situated close to the dorsal surface of the abdomen. Use a hand lens to see the series of small pores or ostia situated on the swelling on the heart.

4. The internal cavity is almost filled with the digestive tract. There is a short esophagus leading from the mouth into a large crop which fills most of the thorax. A small gizzard joins with the stomach. Where the gizzard and stomach meet there are 6 conical, thin-walled pouches, called the gastric caeca, which secrete the digestive enzymes. At the posterior end of the stomach there is a tangled mass of threads that are the Malpighian tubules. These tubules act as kidneys to elimate the nitrogenous wastes from the blood. The intestine extends from the stomach as a narrow tube until it widens to form the rectum. The anus is a posterior dorsal opening.

5. Locate the ventral (bottom) nerve chord. Trace it to the head region and at the same time locate several of the swollen regions of the cord, or ganglia.

6. With the aid of a hand lens, locate some of the tracheae or breathing tubes which lead to the external environment by means of breathing pores or spiracles. With scissors, make a careful incision completely around the spiracle so as to remove it intact from the body wall. Make a wet mount and examine with a compound microscope.

DISCUSSION:

1. Does the grasshopper have a well-defined head?
2. How many segments are there in the thorax?
3. To which thoracic segments are the wings attached?
4. Which body region bears the legs?
5. What is the function of the antennae? How does their structure enable them to carry out this function?
6. How is the structure of the grasshopper's compound eye related to its function?

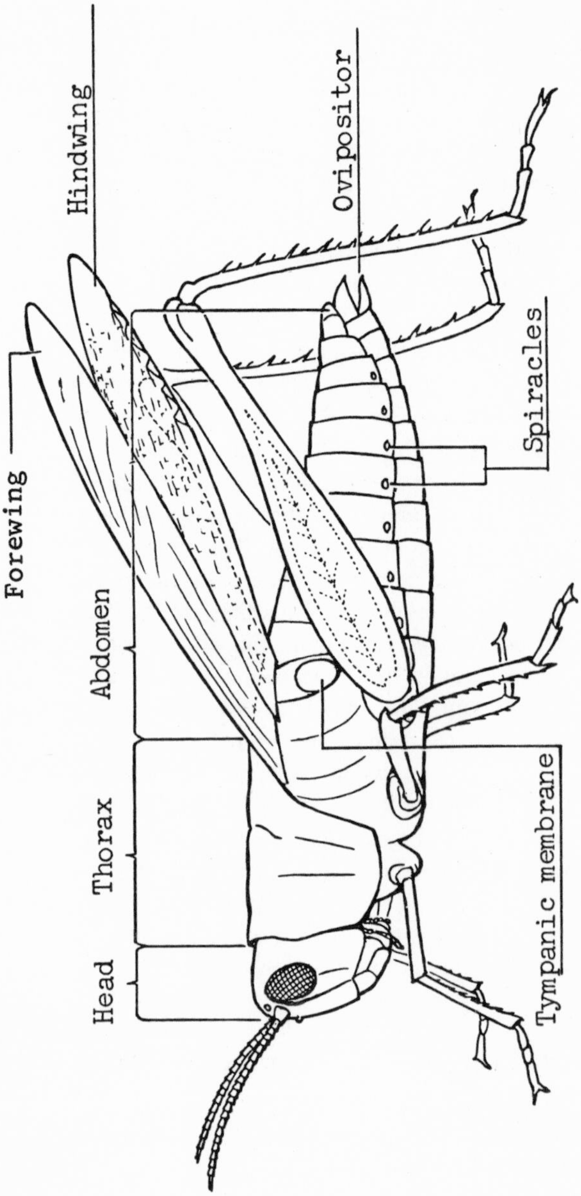

Diagram 23. External Anatomy of the Grasshopper.

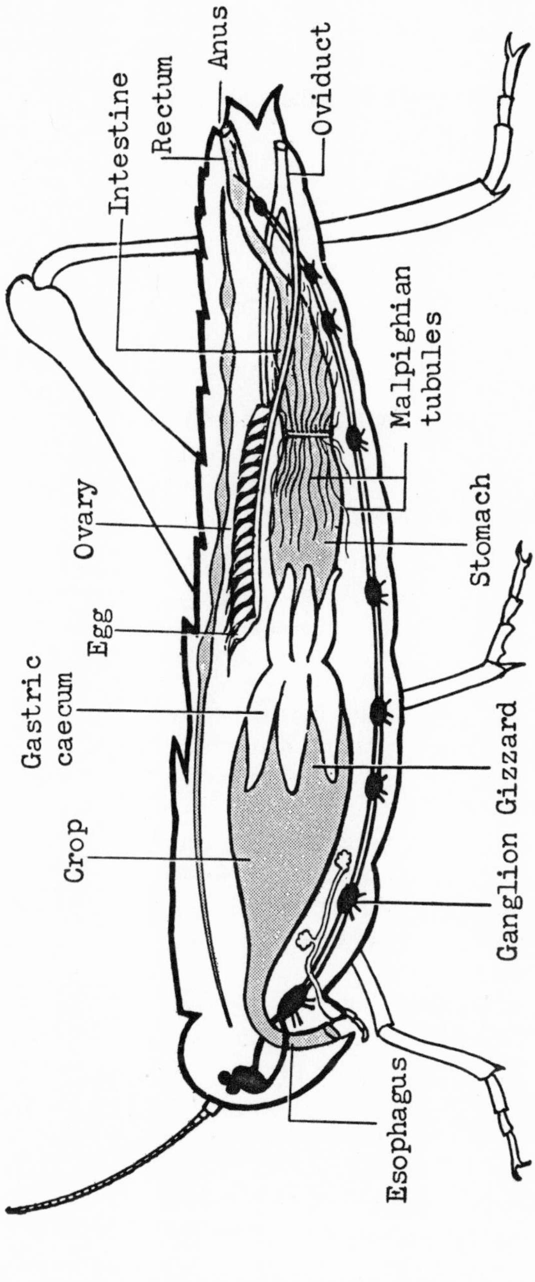

Diagram 24. Internal Structure of the Grasshopper.

7. How is the structure of the grasshopper's hind legs related to their function?

8. What adaptations are to be found on the maxillae and labium which aid in food getting?

9. How is the basic structure of the grasshopper and earthworm similar? Different?

THE FROG

AIM: To study the internal structure of the frog.

INTRODUCTION: Since the frog is a chordate, we study it as an example of the chordates. At the same time it is an Amphibian and as such has certain organs that differ from those of man. In this exercise, you will note the differences as well as the similarities.

MATERIALS NEEDED: Preserved frogs (we use preserved frogs since we have found that they are easier and safer to handle when used in quantity), dissecting pan, scissors, 2 dissecting needles, pins, forceps.

PROCEDURE:

1. Although the frogs should be thoroughly washed in running water prior to their use in the laboratory, it is advisable that each student rinse the frog some more so that it will be easier to handle and the preservative will not be too noticeable.

2. Place the frog on its back in the dissecting pan. Pin the legs down spread-eagle with the aid of several pins. Now use the forceps and scissors and cut through the skin along the midline by lifting the skin with the forceps and starting the cut at that time. Pin down the flaps of the skin. What layer has now been exposed?

3. Cut through the muscle layer of the abdomen and proceed anteriorly through the pectoral girdle, following a line a little to the side of the midline. (CAUTION: When using the preserved frogs, do not permit formaldehyde to squirt into your eyes.) Be careful to avoid cutting into the underlying organs. Pin down the abdominal muscles so as to expose the cavity.

4. You may now see a large black and white mass. This is the eggs within the ovaries. Lift them up and remove them very carefully so that no other body tissue is removed along with the eggs. Explain the absence of such eggs in some frogs.

5. Locate the organs indicated below; remove them to a piece of notebook paper and label:

   a. a lung

b. the heart
c. a lobe of liver
d. gall bladder
e. stomach
f. spleen
g. intestine
h. ovary or testis
i. a kidney

6. Remove all the abdominal organs and find the spinal cord and spinal nerves.
7. Open the frog's mouth. How does the tongue differ from your tongue?
8. Find the sciatic nerve in a leg.

DRAWING:

Label Diagram 25.

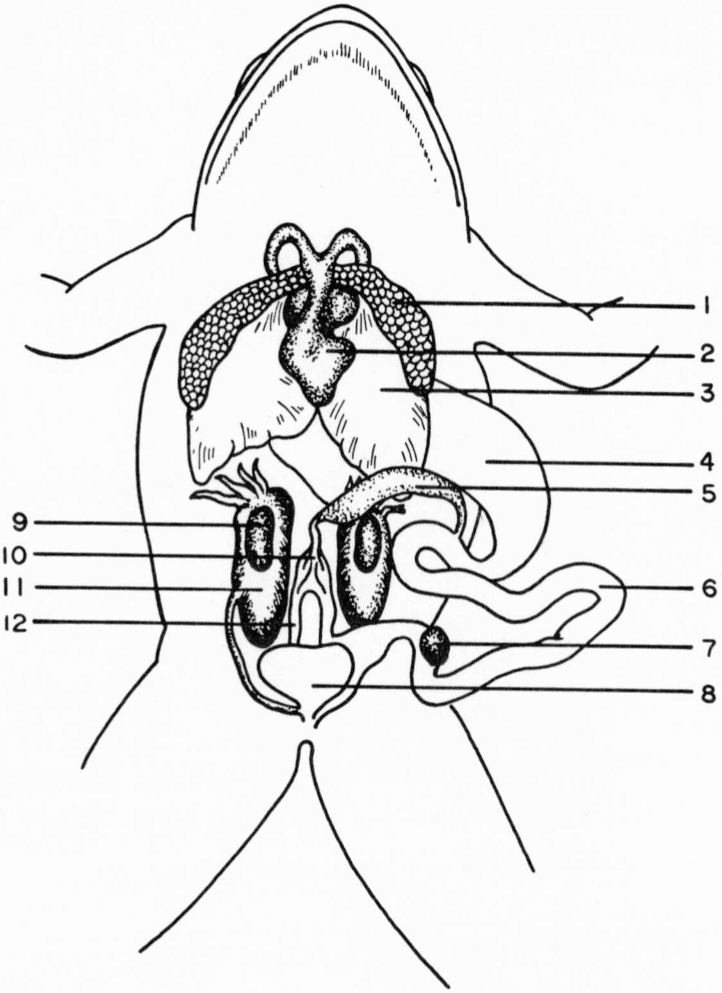

1 ______________________ 7 ______________________

2 ______________________ 8 ______________________

3 ______________________ 9 ______________________

4 ______________________ 10 ______________________

5 ______________________ 11 ______________________

6 ______________________ 12 ______________________

Diagram 25.

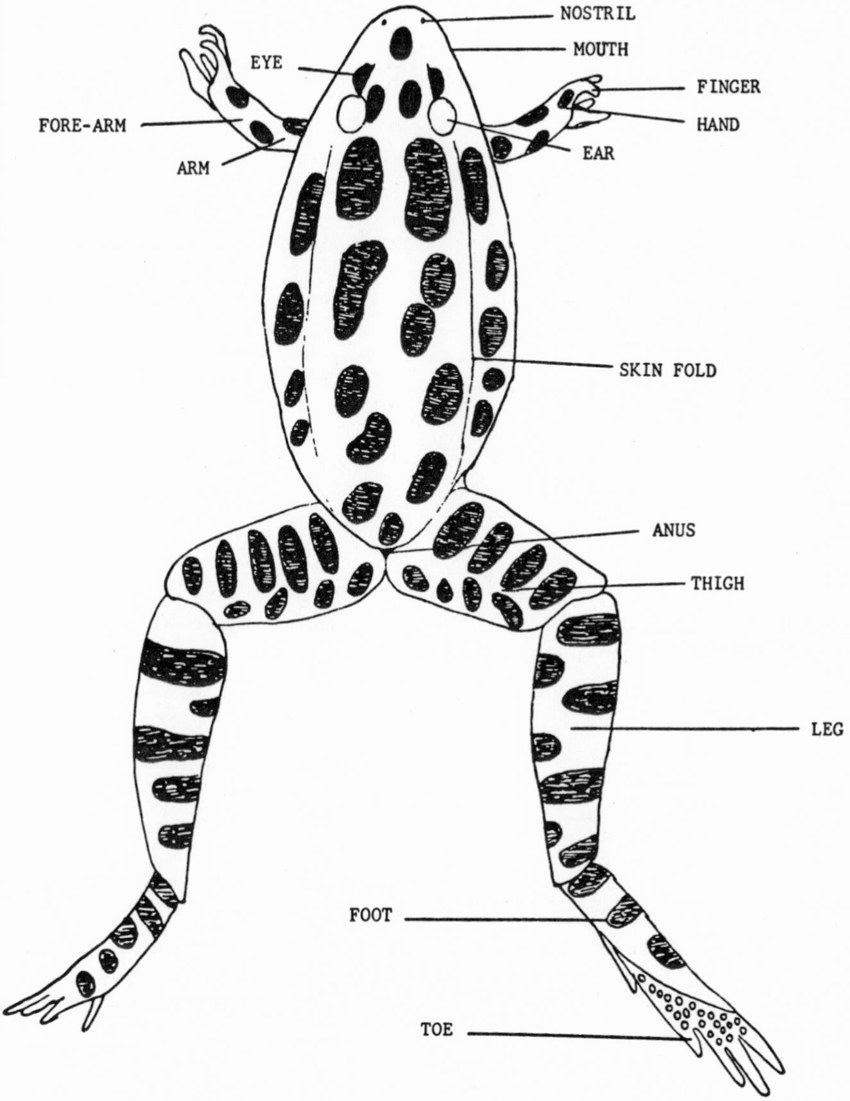

Diagram 26. External Parts of the Frog.

UNIT IV

MAINTENANCE IN MAN

I. NUTRITION

A. Human Digestion

1. *Digestive System*: The human digestive system consists of a continuous tube, the alimentary canal, and the accessory organs which function in conjunction with the alimentary canal.

a. The alimentary canal: The alimentary canal consists of the following specialized subdivisions:

- oral cavity
- pharynx
- esophagus
- stomach
- small intestine
- large intestine

b. Digestive glands: Digestive glands occur as specialized groups of cells lining the alimentary canal or in separate, accessory organs. Enzymes are secreted into the digestive tube where extracellular digestion occurs.

- Glands within the digestive tube include the gastric glands of the stomach wall and the intestinal glands in the lining of the small intestine.
- Accessory glands include the salivary glands, and the glands of the liver and pancreas. Secretions of these glands reach the digestive tube through ducts.

2. *Digestive Function*: Mechanical digestion results from the action of the teeth and tongue in the oral cavity and the muscular contractions

(peristalsis) in the esophagus, stomach, and small intestine.

Chemical digestion begins in the oral cavity and is completed in the small intestine.

- Carbohydrate digestion begins in the oral cavity and is completed in the small intestine.
- Protein digestion begins in the stomach and is completed in the small intestine.
- Lipid digestion occurs in the small intestine.

Undigested and indigestible materials accumulate in the large intestine where water is absorbed into the blood. The remaining materials form feces which are periodically evacuated through the anus.

II. TRANSPORT

A. Circulation:

1. Circulatory system

a. Structures

(1) Heart: The human heart is a pumping device which consists of four chambers—two collecting chambers, the atria, and two pumping chambers, the ventricles.

(2) Arteries: Arteries are relatively thick-walled, muscular vessels which transport blood from the heart to parts of the body. All arteries lack valves.

(3) Capillaries: Capillaries are blood vessels with walls only a cell thick. This feature permits ready exchange of materials between the blood and cells adjacent to the capillaries. Intercellular fluid (ICF), which surrounds all body cells, permits transport of materials to and from cells which are remote from capillaries.

(4) Veins: Veins are relatively thin-walled vessels which transport blood from capillary beds back toward the heart. Many veins contain valves which prevent back-flow of blood in the system.

(5) Lymph Vessels: Lymph vessels make up a system of tubes which are essentially independent of the blood transport system. Lymph vessels transport lymph. The lymph and blood systems connect at one site where lymph empties into the blood.

b. Fluids:

(1) Blood: Blood consists of a fluid (plasma) in which are suspended red and white blood cells and platelets. The plasma contains water, ions, and dissolved or suspended compounds including blood proteins, gases, and digestive and products.

(2) Lymph: All cells of the body are bathed in a watery fluid (ICF) which originates from blood plasma. Known as lymph when it is absorbed into lymph vessels, the intercellular fluid consists mostly of water and small amounts of proteins, salts, and all the parts of lymph.

2. Circulatory function:

a. Transport: The principal function of the human circulatory system is transport of dissolved and suspended materials throughout the body.

b. Protection: The circulatory system contains mechanisms which have a protective function, including protection from blood loss through injury and from the harmful effects of disease-causing organisms (pathogens).

(1) Clotting: The clotting of blood is an example of an enzyme-controlled activity. A clot forms when fibrin, a protein material consisting of long fibrous strands, is synthesized. The synthesis of fibrin results from enzyme-regulated reactions among disintegrated blood platelets, blood proteins, and calcium ions in the plasma.

Fibrin tends to trap blood cells and fluid, and forms a mat-like mass which prevents blood loss from a wound.

(2) Phagocytosis: Some white blood cells exhibit the property of engulfing bacteria and other pathogens. Such cells are known as phagocytes ("eating cells") and their action is known as phagocytosis.

(3) Immunological Reactions: Certain blood proteins belong to the class of compounds known as antibodies. Antibodies are synthesized when foreign proteins (antigens) are introduced into the body. Antigen-antibody reactions tend to destroy, or neutralize, the harmful effects of antigens.

III. RESPIRATION

A. *Respiratory System*: The lungs are the principal respiratory organs in man. A system of passageways permits atmospheric air to enter and leave the lungs as a result of rhythmic contractions of the diaphragm and the muscles of the rib cage.

The air passageways include the oral cavity, nasal passages, pharynx, larynx, trachea, bronchi, bronchial tubes, and alveoli.

B. *Respiratory Function*: In the alveoli some of the oxygen from the air diffuses into the blood, as carbon dioxide diffuses into the air space. Capillaries which surround the alveoli permit this gas exchange which is referred to as external respiration.

Breathing, the alternate expansion and contraction of the lungs, serves to replenish oxygen-rich atmospheric air in the alveoli. The breathing rate is controlled by the nervous system and varies with the concentration of carbon dioxide in the blood.

The transfer of oxygen and carbon dioxide between capillaries and body cells is referred to as internal respiration.

Cellular respiration is a complex process involving the energy transfer required for maintaining life.

IV. EXCRETION

A. *Excretory System*: The organs of the human excretory system include the lungs, liver, skin, kidneys, and rectum.

B. *Excretory Function*:

1. Lungs: The diffusion of carbon dioxide into the alveoli of the lungs is the principal means of carbon dioxide excretion in man. In addition, water, in the form of vapor, is excreted in exhaled air.

2. Liver: The breakdown of blood cells and other materials results in a large amount of urea synthesis in the liver. The urea is absorbed into the blood and transported to the kidneys.

Bile salts are discharged into the small intestine from where they move to the large intestine for elimination with fecal material.

3. Skin: The sweat glands of the skin permit the excretion of water and dissolved salts onto the surface of the skin. Water evaporates and salt crystals are removed by physical contact with clothing and other surfaces.

Perspiration, in addition to serving as an excretory fluid, is a major temperature-regulating device. Evaporation of perspiration utilizes heat from skin cells, thus lowering their temperature.

4. Kidney: The functional unit of the human kidney is the nephron. The nephron consists of a long, coiled tubule with a cup-shaped capsule at one end.

Capillary beds within the capsules discharge water, salts, urea, amino acids, and glucose into the nephron. As these materials move through the tubule of the nephron, water, minerals, and digestive end products are reabsorbed by active transport into capillaries associated with the tubule.

After reabsorption, the fluid that remains consists mostly of water, urea, and salts, and is known as urine.

Urine flows from the kidney through the ureter to the bladder. Periodically, urine is discharged from the bladder through the urethra.

V. SYNTHESIS:

Synthesis occurs throughout the entire body. As a result, there is no "synthesis system" as such.

A. *Limitations*: Humans produce enzymes which permit the synthesis of most of the compounds necessary for normal physiological activity. However, man cannot synthesize some substances from available raw materials. These include certain amino acids and vitamins.

B. Synthesis in relation to diet: The amino acids and vitamins man cannot synthesize are synthesized by a wide variety of plants and other animals.

Since a well-balanced diet will include necessary substances, food supplements are rarely necessary.

VI. REGULATION

A. NERVOUS REGULATION

1. *Nervous System*: The behavior of man is dependent upon, and integral with, his complex nervous system.

a. Structural units: The varied aggregations comprise differing organizations of three basic neuron types.

(1) Sensory neurons: Sensory neurons transmit impulses from sense organs (receptors) to the central nervous system.

Sense organs, where major concentration of sensory neurons occurs, include the eyes, ears, tongue, nasal passages, skin, and muscles.

(2) Associative neurons: Associative neurons serve to interpret stimuli, to orginate impulses to motor neurons, and to relay impulses between sensory and motor neurons.

(3) Motor neurons: Motor neurons transmit impulses from the central nervous system to effectors (muscles and glands).

b. Organization: The nervous system consists of several different organizations of neurons which are distinguishable on the basis of structure and function.

(1) Nerves: Nerves are bundles of neurons (sensory, motor, or both) which are specialized for the transmission of impulses over long distances.

(2) Ganglia and plexuses: Ganglia are enlarged structures which contain associative neurons and serve as relay centers, redirecting incoming impulses toward nerve centers or effectors.

Plexuses resemble ganglia in their function but are larger, sheet-like masses of nervous tissue.

(3) Spinal cord: The spinal cord with its spinal nerves is an extension of the brain. It has coordinating functions, is a center for many reflex actions, and serves to conduct impulses between the brain and other body structures.

(4) Brain: The brain is a large mass of neurons organized in three major divisions, each with specialized functions.

(a) Cerebrum: In the cerebrum, sensory impulses become sensations, motor activities frequently are initiated, and conscious acts have their origin. The cerebrum functions as a memory bank and an integration center capable of creative thought.

(b) Cerebellum: In the cerebellum, motor patterns are coordinated so that the organisms respond in an integrated manner rather than with random movements.

(c) Medulla: Nerve impulses controlling the involuntary body activities such as breathing, heartbeat, blood pressure, movements of the digestive system, coughing, and sneezing, originate in the medulla.

2. *Nervous Functions*:

a. Involuntary behavior: Involuntary behavior is automatic and may be either inborn or acquired.

(1) Reflexes: Reflex actions are inborn, automatic patterns of behavior which are important in controlling everyday vital functions.

Reflex behavior involves a characteristic route (reflex arc) over which impulses travel. In a spinal reflex there is a pathway from (a) sense organ (receptor) to (b) sensory neuron to (c) associative neurons in the spinal cord to (d) motor neuron to (e) muscle or gland (effector).

(2) Conditioned behavior: Conditioned behavior is automatic behavior which is acquired. Habits are a form of conditioned behavior which involve the repeated performance of an act. This establishes pathways of nerve impulse transmission which permit automatic responses.

b. Voluntary behavior: Voluntary behavior involves the cerebrum of the brain. It incorporates memory of past experiences, association, and judgment. Voluntary behavior is better developed in man than in other animals.

B. ENDOCRINE REGULATION

1. *Endocrine system*: The endocrine system consists of the endocrine glands and their secretion. Endocrine glands include: the thyroid gland, the parathyroid glands, adrenal glands, the pituitary gland, the pancreas, and the sex glands.

2. *Endocrine Function*:

a. Thyroid gland: The thyroid hormone, thyroxin, regulates the rate of oxidation in the body and is essential for normal mental and physical development.

b. Parathyroid glands: Parathormone, the secretion of the parathyroids, controls the metabolism of calcium which is necessary for nerve function, blood clotting, and proper growth of teeth and bones.

c. Adrenal glands: The adrenal glands secrete a variety of hormones including adrenalin (epinephrine), cortisone, and cortin.

Adrenalin regulates rate of heartbeat, blood sugar levels, blood-clotting rate, and so forth.

Cortisone is necessary for the healthy condition of cartilage in joints at the ends of bones.

Cortin controls the levels of salts in blood, and the blood pressure.

d. Pituitary gland: The pituitary gland secretes a variety of hormones, many of which influence the activity of other endocrine glands. Examples include:

- ACTH which stimulates the adrenal glands to produce cortisone.
- FSH which stimulates activity in the ovaries.

e. Pancreas: Located in the pancreas are small groups of cells (the islets of Langerhans) which secrete the hormones of insulin and glucagon.

Insulin regulates blood sugar levels by promoting the outflow of sugar from the blood, e.g., into the storage areas of the liver and muscle.

Glucagon stimulates the discharge of sugar from the liver into the blood.

f. Sex Glands: Male sex glands, located in the testes, secrete testosterone, the hormone which influences male secondary sex characteristics.

Female sex glands secrete estrogen and other hormones. Estrogen influences female secondary sex characteristics.

VII. LOCOMOTION

A. *Locomotive system*

1. Bones: The human skeleton consists of bones of various sizes and shapes. The function of these bones includes:
   - support of body structures
   - serving as anchor sites for muscle action
   - serving as levers for body movement
   - serving as protection for delicate internal organs.

2. Muscles: The voluntary muscles, which control movements of the skeleton, are of two types. Extensors serve to extend the limbs. Flexors act to return extended limbs closer to the body. Movements of the skeleton involve the coordinated interaction of flexors and extensors working in pairs.

Other muscles, not involved in locomotion, include:

- smooth muscles which function in involuntary muscle activity such as breathing, peristalsis, and so forth.
- cardiac tissue which is found in the heart and maintains its beating action.

3. Tendons: Tendons are composed of connective tissue and serve to join muscles to bones.

4. Ligaments: Ligaments also consist of connective tissue and bind ends of bones together forming joints, such as those of the elbow or knee.

B. *Locomotive function*: The interaction of paired muscles serves to move the long bones of the skeleton and produce locomotion. Coordination of the muscles and bones is achieved by the nervous system and permits walking, running, and so forth.

SALIVARY ENZYME ACTIVITY AND pH*

AIM: To determine which pH is best for the salivary enzyme amylase to function at its maximum.

As you know acids donate hydrogen ions while bases are the substances which combine with hydrogen ions. You also know that the concentration of hydrogen ions in an aqueous solution determines the degree of acidity of the solution. The term pH is used to express the hydrogen ion concentration of a solution. A pH of 7 is neutral—neither acid nor base. Anything above 7 is basic and anything with a pH under 7 is acidic.

Cell metabolism and enzymes depend upon the proper pH. Just a slight

* Starred experiments throughout the book should be checked for materials at least two days in advance. Some materials must be prepared ahead of time.

variation from the necessary pH will slow down the reaction considerably.

In this exercise, you will determine what the pH will be for the maximum efficiency of amylase in saliva to start the breakdown of starch to maltose.

MATERIALS NEEDED: Test tubes test tube rack; 25 ml graduate cylinders; small beakers; spot plates; medicine droppers; flasks; starch solution; iodine solution; buffer solution of pH 3, 5, 7, and 9; saliva (containing amylase); cheesecloth; marking pencil; sterile rubber bands or sterile paraffin; Benedict's solution; Bunsen burner; test tube holder.

PROCEDURE:

1. Chew on rubber bands or a piece of paraffin until the saliva flows freely. Collect about 1/4 of a test tube full. Add an equal amount of water.

2. Shake the test tube thoroughly to mix and then filter through a double thickness of cheesecloth into a beaker.

3. Make up the pH solution of 3, 5, 7, and 9 by using the buffer tablets —one dissolved in 100 ml of distilled water. You can speed up the dissolving by crushing the tablet with a mortar and pestle.

4. When the buffer solutions are ready, take 5 ml of each and place each into a test tube. Label each test tube according to the pH of the solution. The buffer solution is a mixture of either a weak acid or base with one of its salts which maintains the pH of the solution in which it has been placed.

5. Add 5 ml of starch solution to each of the test tubes and shake thoroughly to mix the contents. The starch is made by adding 5 gm of a soluble starch to 100 ml water. Boil and stir thoroughly to dissolve the starch. Allow to cool before using.

6. Add 5 drops of the salivary solution to each tube and shake thoroughly. Test each tube for the presence of starch by taking a few drops of the starch-saliva solution and testing with the iodine solution.

7. Continue to test the contents of each tube at one-minute intervals. As the starch changes to dextrin, the color becomes brown. When it remains colorless, it means that the substance now in the test tube is maltose.

8. To check the presence of the maltose, use the Benedict's solution for a sugar test. A red-brown precipitate means sugar is present.

9. You should find that the starch is best changed into dextrins and then sugar at a pH of 7, with the next best being a pH of 5. The pH of 3 still shows starch present at the end of 15 minutes as it also does at a pH of 9.

CONCLUSIONS AND DISCUSSIONS:

1. In light of the above findings, why do you think that amylase would be able to function in the stomach? in the small intestine?

2. In light of the above findings, what would you say the pH of your saliva is? How can you prove this?

GASTRIC DIGESTION*

AIM: What is the effect of pH upon the digestion of protein by pepsin?

It is necessary that there be a specific pH in the stomach for proper digestion. In this exercise, we shall attempt to find out what this pH should be. We are using egg white coagulum for the protein.

MATERIALS NEEDED:

At the front desk: fingerbowl containing raw egg white; capillary tubing, 6 pieces each about 8 inches long; Erlenmeyer flask or pot, 8 inches deep; tripod; Bunsen burner; 4 triangular files and ampule knives.

For Each Group of Students: Test tube rack or glass tumbler for holding test tubes; 4 test tubes; 25 ml graduate; labels; bottles, each containing as follows: (1) HCl, 0.8%, (2) Pepsin, 2.0%, (3) $NaHCO_3$, 0.8%, (4) water.

PROCEDURE:

(A) Preparation of Solution:

Each member of each group labels the four test tubes by recording the class, group, and tube number, e.g. Bio 106, 3, IV.

A different student in each group prepares each of the four solutions. Wash the graduate each time it is used.

	Solution	*Preparation*	
I.	Acid (0.4% HCl)	HCl (0.8%)	5 ml
		Water	5 ml
II.	Pepsin (1.0%)	Pepsin (2.0%)	5 ml
		Water	5 ml
III.	Pepsin (1.0%)	Pepsin (2.0%)	5 ml
	HCl (0.4%)	HCl (0.8%)	5 ml
IV.	Pepsin (1.0%)	Pepsin (2.0%)	5 ml
	$NaHCO_3$ (0.4%)	$NaHCO_3$	5 ml

(B) Preparation of Egg-White Tubes

The protein used is coagulated egg white, prepared from a fresh egg. One member of each group goes to the front demonstration table and draws egg white into a piece of glass tubing about 8″ long. The albumin is then coagulated by placing the tubes into water at 85°C, for 5 minutes. The tubes are then cut into sections about 1″ long. Each section is placed into each of the four test tubes, as shown in Diagram 27, which meanwhile have been prepared by other students in the group.

(C) Incubation

Place all tubes into a glass tumbler which will fit into the incubator. Incubate for 24 hours at 37°C

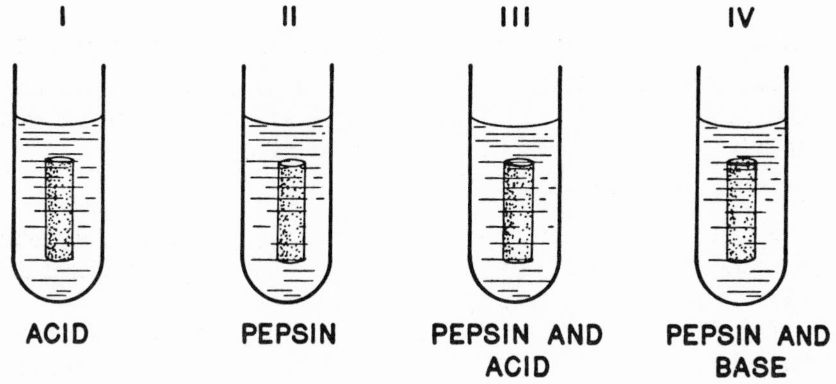

Diagram 27.

OBSERVATIONS:

The tumblers containing the tubes should be examined the following day. Test tube number III should show the most digestion. Use a millimeter ruler to make any measurements that will be of significance in your findings. Make drawings.

EXERCISES:

Write a report of the exercise in which should be included:

A. Aim

B. Method. Indicate the main steps of the procedure but refer to direction sheets for details such as how the solutions were prepared.

C. Observations. Include a drawing.

D. Discuss. Answer the following questions:

(1) What is a protein?

(2) What is an enzyme?

(3) How do chemists use pH to indicate whether a solution is acid or alkaline?

(4) What is the effect of pH upon gastric protease?

(5) How does the stomach's acidity affect the action of ptyalin?

(6) To prepare an "artificial gastric juice," what substances would you use?

(7) See what you can learn about ninhydrin. What are proteins changed to during digestion? How can this be verified?

(8) Criticize this experiment from the viewpoint of controls.

E. Conclusion.

These directions are adapted from a laboratory exercise which has been done at Far Rockaway High School, Far Rockaway, N.Y.

DIGESTION IN THE SMALL INTESTINE*

AIM: To learn what digestive enzymes are found in the small intestine.

The final steps in digestion are carried out in the small intestine. Fats are changed into fatty acids and proteins into amino acids. Pancreatin contains the digestive enzymes used in these changes.

MATERIALS NEEDED: Pancreatin powder, chopped hard-boiled egg, 11 test tubes, phenolphthalein solution, 0.5% sodium carbonate solution, 2% boric acid solution, water, olive oil, soap solution or 5% bile salt solution (sodium taurocholate), dried milk powder, litmus powder.

PROCEDURE:

(A) Digestion by Trypsin:

1. Put some chopped hard-boiled egg white into 6 test tubes; then add a pinch of pancreatin along with 10 ml of water.
2. Set 2 tubes aside.
3. To 2 others, add 2 drops of phenolphthalein solution. Add 0.5% solution of sodium carbonate, drop by drop until the first pink color appears. The pH of these tubes will be approximately 8 or on the alkaline side.
4. To the last 2 tubes add 5 ml of a 2% boric acid solution so that they will contain a pH of close to 5. Place all the test tubes into a water bath at 40°C or an incubator for 24 hours. The most digestion will be found in those tubes that were slightly alkaline along with the pancreatin.

(B) Formation of Acid: Pancreatin contains digestive enzymes which split fats into fatty acids and proteins into amino acids. In order to show this reaction, litmus is used as an indicator to show that there are acids formed after digestion.

1. Prepare litmus milk powder by adding 1 part of litmus powder to 40 parts of dried milk powder. Then dissolve one part of the prepared litmus powder in 9 parts of water.
2. In each of 2 test tubes place 10 ml of the solution. Then place a pinch of pancreatin in one, and to the other add only water. Place the tubes in a water bath.
3. Shortly there will be a change of the litmus solution to colorless. This indicates that the milk has broken down to an acid.

(C) Breakdown of Fats:

1. Prepare 3 test tubes containing 10 ml of water and a few drops of olive oil.

2. Add 2 ml of a soap solution to two of them.

3. Add a pinch of pancreatin powder to 10 ml of water and take 5 ml of this solution and place it in one of the test tubes with the soap solution.

4. You now have 3 test tubes: (a) a tube containing water, olive oil, soap solution and pancreatin solution, (b) a control containing water, olive oil, and soap solution, (c) another control containing water, olive oil and pancreatin water.

5. Use litmus paper to show that the solutions are neutral to start with. Place the test into a water bath for 30 minutes.

6. Tube (a) will have acid in it when tested with the litmus after 30 minutes.

CONCLUSION AND DISCUSSION:

The various foods are thus broken down completely and the end products that are useful to the body will be absorbed and sent to various parts for use.

Answer the following questions:

1. Why is there more digestion of the egg when there is a basic solution along with the pancreatin?

2. Why is a soap solution or bile salt solution used in the digestion of fats?

3. What acids would be formed as milk is digested?

KIDNEY FUNCTION

AIM: To learn how the kidney maintains the homeostatic condition of the blood and what is the role of the kidney in excretion.

Excretory organs are found in those animals that consume food containing large volumes of nitrogenous compounds or have skin or scales or an exoskeleton that covers the living cells of the body surface so that the surface cannot act as an excretory surface.

In the human, the kidney serves another purpose besides excretion. It is also used to maintain a constant level of chemical substances in the blood.

In order to show what is the effect of food intake on the composition of the urine, it is necessary that different students have different diets for one day prior to the laboratory test.

There are 5 different diets:

Diet 1. High Protein: Meat and other high-protein foods are eaten while starches, fats, and sugars are not.

Diet 2. Low-salt: Drink plenty of water and make certain that the diet contains no salt of any kind.

Diet 3. High Glucose: Eat more candy bars and less regular food. Chocolate cakes and cookies are also on the diet.

Diet 4. High Pentose: Eat fruits such as plums, pears, and bananas, instead of most of your regular diet.

Diet 5. Regular diet: Eat what you usually eat (which probably contains a little of everything).

MATERIALS NEEDED: A stoppered bottle for urine sample, 50 test tubes and corks, glacial acetic acid, a methyl alcohol solution of xanthydrol, pipette, 10% silver nitrate solution, Clinitest paper, Bunsen burner, large beaker, graduated cylinder, Bial's reagent, 1.0% ferric chloride solution, 0.1M urea solution, 0.1% glucose solution, 0.1% sodium chloride solution, 0.25% pentose sugar solution.

PROCEDURE:

The class should work in groups of 5, each member of the group choosing a different diet so that the entire set of tests can be done by one group and the information compared and exchanged.

(1) Each student should adhere to his particular diet for the day previous to the laboratory. On the morning of the laboratory, the student should collect his urine sample immediately upon awakening and label the bottle with the proper diet number

(2) Each student will test his own urine for all 4 tests, as follows:

a. *Test 1.* Urea: Dilute 1 ml of urine with water to make 500 ml of solution. Transfer 5 ml of this diluted urine to a test tube and add 5 ml of glacial acetic acid and 0.5 ml of a methyl alcohol solution of xanthydrol. Cork the tube and shake it vigorously. Into a second tube, place 5 ml of the urea solution found on your desk and add 5 ml of glacial acetic acid and 0.5 ml of a methyl alcohol solution of xanthydrol. Cork the tube and shake it vigorously. Mark the tubes 1a for the urine and 1b for the urea tube which is a control. Allow them to stand for at least 1 hour and overnight if necessary. If there are large, loose clumps in the tubes, urea was present.

b. *Test 2.* Chloride. Dilute 1 ml of urine with water to make 500 ml. Transfer 5 ml of diluted urine to a test tube and add a few drops of 10% silver nitrate solution. To another tube add 0.1% sodium chloride solution plus a few drops of 10% silver nitrate solution. The second tube is the control and will show a cloudy white precipitate indicating that a chloride is present. If there is a precipitate in the urine sample then there is chloride present in the sample.

c. *Test 3.* Glucose. Dip a piece of Clinitest paper in the urine sample. Then compare the color of the paper with the chart provided with the paper to obtain the approximate percentage of glucose in the urine. Dip the paper into the glucose solution to compare the results of both.

d. *Test 4.* Pentose sugar. To 2 ml of urine in a test tube and to 2 ml of 0.25% pentose sugar solution in another test tube add 2 ml of Bial's reagent. (Note: Corrosive. Be careful not to get any on your skin. If you do, wash under heavily running water.) Gently warm the tubes over the Bunsen burner and then add a few drops of the 1.0% ferric chloride (also corrosive).

(3) The results should be tabulated on a separate sheet of paper and should generally show the following:

a. A high-protein diet will increase the amount of urea in the urine.

b. A salt-free diet should lower the amount of chloride in the urine.

c. A high-glucose diet may increase the amount of glucose in the urine.

d. A high-pentose diet will increase the amount of pentose in the urine.

e. A regular diet should not show any unusual amounts of any one substance in the urine.

CONCLUSIONS:

We thus see that the amount of the various substances found in the urine depends upon the substances in the foods eaten. The kidney thus serves to maintain the homeostatic condition of the blood, selectively removing the excess substances from the blood and excreting them. Without the kidney functioning as it does there would be a widely varied amount of dissolved substances in the blood which would result in all sorts of physical malfunctions and illnesses.

Do a report on various kidney malfunctions.

MEASUREMENT OF EXHALED CARBON DIOXIDE

AIM: To determine the quantity of carbon dioxide in your exhaled breath and to make comparisons with other members of the class.

INTRODUCTION: In this exercise you will make measurements pertaining to the carbon dioxide in the breath you exhale. The general procedure is to breathe into water where the CO_2 forms carbonic acid in accordance with the equation:

$$CO_2 + H_2O \rightleftarrows H_2CO_3 \qquad (1)$$

The amount of H_2CO_3 produced is determined by finding out how much of a known strength of NaOH is required to neutralize the acid, as shown by the equation:

$$NaOH + H_2CO_3 \longrightarrow NaHCO_3 + H_2O \qquad (2)$$

An overall equation would be:

$$\underset{40}{NaOH} + \underset{44}{CO_2} + \underset{18}{H_2O} \longrightarrow NaHCO_3 \qquad (3)$$

The indicator phenolphthalein will be used which is pink in the basic range and colorless at other ranges. The process of adding an acid or base to a solution until the indicator changes color is known as *titration.*

Under equation (3) are listed the molecular weights which show the ratios by which these substances combine in this reaction. These molecular weights are *relative* weights and could be expressed in tons, pounds, or ounces. When the molecular weights are taken in *grams* the weight is called a mole. A mole of NaOH is *40 grams*, a mole of CO_2 is *44 grams.* However, since 44 grams of CO_2 is too large a unit for use in this exercise, measurements will be made in *micromoles.* A micromole is a *millionth* of a mole or 10^{-6} moles.

The NaOH used for titration has been prepared at a concentration of 0.04% or 0.01M. This has been so calculated that 1 ml of the NaOH neutralizes or combines with 10 micromoles of CO_2. Thus, the number of ml of NaOH required to neutralize the H_2CO_3 multiplied by 10 is equal to the micromoles of CO_2 present.

Statistical comparisons of data will be made from the class as a whole.

MATERIALS: Burette attached to a ring stand, wide-mouth bottle, 0.04% NaOH solution, phenolphthalein indicator, timer, straws to blow through into the water, water.

PROCEDURE:

1. To 100 ml of tap water in a bottle add 5 drops of phenolphthalein indicator. If the solution is colorless, add a few drops of 0.04% NaOH solution, and swirl, to make the solution slightly pink (alkaline). CAUTION: NaOH is a strong alkali. Do not permit contact with the skin or clothing. In the event of accident flush with much water.
2. Blow through a tube into the water for *exactly one minute.* Inhale in normal fashion but exhale into the tube. Start measuring the time with an inhalation. Avoid blowing so hard as to splash water from the bottle. There will be a color change as the water changes to carbonic acid. What is the color change?
3. Introduce about 20 ml of the 0.04% NaOH into the burette and place the bottle below the burette. Record the level of the NaOH at the start. Perform the titration by adding the NaOH slowly, a drop at a time, to the bottle. Swirl the bottle as you do so. Add sufficient NaOH until the pink color lasts for one minute. Record the level of NaOH in the burette upon completion. Determine the number of ml of NaOH required.

DATA AND CALCULATION:

Level of NaOH in burette at start . . .ml
Level of NaOH in burette at end . . .ml
Volume of NaOH required in titration . . .ml

Calculate the number of micromoles of CO_2 present in the water solution.
. . .micromoles

Record this result on the table for data from the whole class.

CLASS DATA

1. Copy the data for the class:

Pupil	*Micromoles*	*Pupil*	*Micromoles*	*Pupil*	*Micromoles*
1	. . . . .	11	. . . . .	21	. . . . .
2	. . . . .	12	. . . . .	22	. . . . .
3	. . . . .	13	. . . . .	23	. . . . .
4	. . . . .	14	. . . . .	24	. . . . .
5	. . . . .	15	. . . . .	25	. . . . .
6	. . . . .	16	. . . . .	26	. . . . .
7	. . . . .	17	. . . . .	27	. . . . .
8	. . . . .	18	. . . . .	28	. . . . .
9	. . . . .	19	. . . . .	29	. . . . .
10	. . . . .	20	. . . . .	30	. . . . .

2. Prepare a *frequency distribution* for the class.
   Intervals Numbers
3. Which interval was the "most popular"? This is the *mode*. . . . . . .
4. Make a graph of the class data.

DISCUSSION:

1. In this exercise you determined the amount of CO_2 present in the water solution. But this procedure was employed with the aim of determining *the amount of CO_2 you exhale* in one minute. What were the sources of error in the procedure?

2. If the aim had been to determine the *amount of CO_2 your body produces* in one minute, what additional errors would have been present?

3. What factors made for lack of accuracy in the use of the data for comparing the metabolism of the various members of your class?

BLOOD TYPING

AIM: To determine your blood type and to see what happens when blood of different types is mixed.

In 1900 Landsteiner discovered the basic blood types and this led to a great deal of work on blood. Doctors were able to understand why sometimes a blood transfusion was successful while at other times it was not.

Individuals may or may not have one or both of the known clumping factors in their red blood cells. The blood type of an individual will be determined by these clumping factors. There are two types of antibodies found in the blood serum which react against the clumping factors, but in any one person the antibodies of that person will never be antagonistic to his own clumping factors.

Blood type	*Clumping factor in red blood cells*	*Antibodies in blood serum*
A	A	B
B	B	A
AB	AB	NONE
O	NONE	A and B

In this exercise you will be introduced briefly to the topic of blood typing by adding Anti-A serum and Anti-B serum to drops of blood and thus be able to determine the blood type.

MATERIALS: Anti-A serum (colored blue), Anti-B serum (colored yellow), toothpicks, slides, disposable sterile lancets, alcohol, absorbent cotton, microscope slides, glass-marking pencils, 0.85% salt solution and the sera diluted 1/4, 1/16, or oxalated blood.

PROCEDURE:

(A) Typing Your Blood:

1. Mark a slide with a glass crayon as shown in Diagram 28. Place a drop of blood in each circle.

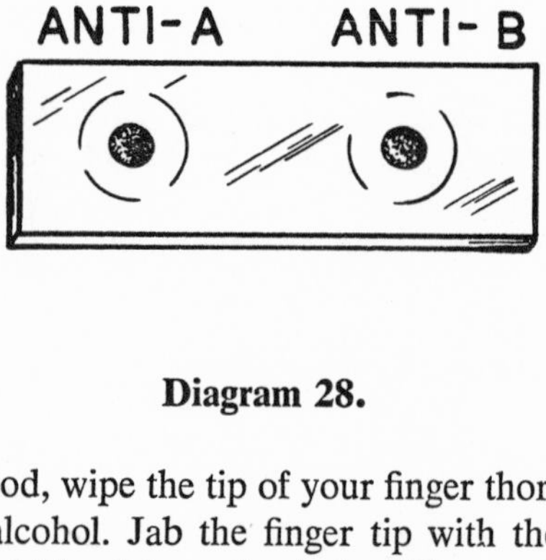

Diagram 28.

2. To draw the blood, wipe the tip of your finger thoroughly with a piece of cotton soaked in alcohol. Jab the finger tip with the sterile lancet and then place the drop of blood in each circle. Wipe your finger thoroughly with the alcohol again.
3. Add one drop of the Anti-A serum (blue) to the A circle and one drop of the Anti-B serum (yellow) to the B circle.
4. Use different clean toothpicks to mix the serum and blood in each circle.
5. Gently tilt the slide occasionally as you watch for 3 to 5 minutes. Look at both circles under the microscope.
6. Watch for clumping. Clumping in the A circle indicates the presence of A agglutinogens and clumping in the B circle indicates the presence of

B agglutinogens. Type A blood has A agglutinogens, type B has B agglutinogens. Type AB has both, and type O has neither.

7. What is your blood type?

(B) Effect of Mixing Various Bloods:

1. Work with one of your classmates on this part. On a clean slide place a drop of the salt solution. On either side of the salt solution you place a drop of your blood while your classmate places a drop of his. Use a clean toothpick to mix the three drops. Slowly tilt slide as before and wait for results.

2. If there is no clumping, it indicates that the two bloods are compatible. If there is clumping it means that the two bloods are incompatible and therefore the blood of one could not be used in a transfusion for the other. However, occasionally blood of the same basic types may be incompatible since there are other factors in the blood that can cause this incompatibility.

(C) Effect of Dilution of Serum:

Make a clean slide as shown in Diagram 29. Place a drop of A or B blood within each circle and add to each circle the corresponding antiserum, diluted to the indicated strength. Use a tooth pick to stir each circle. Tilt and observe for 5 minutes.

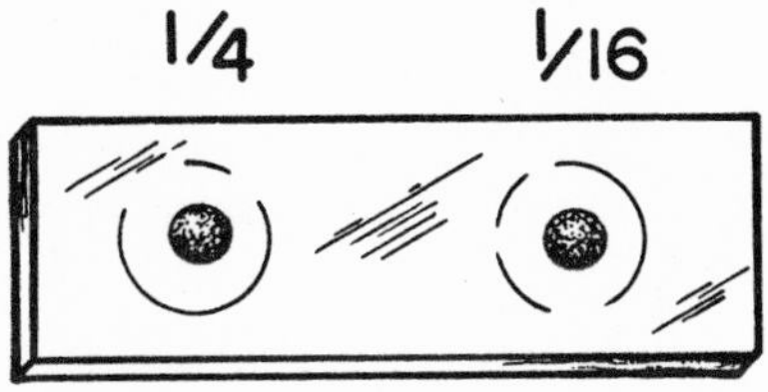

Diagram 29.

The clumping effect will disappear with dilution.

CONCLUSIONS AND DISCUSSIONS:

1. Why is it not necessary to determine the blood type when preparing for a transfusion of blood plasma?

2. Why is type O called the universal donor?

3. What is the difference between blood clumping and blood clotting?

NOTE TO TEACHERS: If the drawing of blood is prohibited by your Board of Education, then you are able to use oxalated blood which can be gotten from a local hospital.

UNIT V

MAINTENANCE IN PLANTS

I. NUTRITION

A. *Autotrophic Nutrition*: Autotrophs are organisms which are able to meet their nutritional needs by the manufacture of their own organic molecules from inorganic materials. Most autotrophs carry on photosynthesis, but a few employ chemosynthetic methods to make food from inorganic sources.

1. Photosynthesis: Photosynthesis is a process in which light energy is converted into the chemical bond energy of organic compounds.

a. Process: About 80 percent of the total photosynthesis on earth occurs in the water, and most of this is carried on by unicellular algae. Photosynthesis is also carried on by multicellular plants. The process is essentially the same in both.

(1) Description: Carbon dioxide and water are chemically utilized inside chloroplasts to form simple sugar molecules (commonly glucose) and oxygen. Light energy is required as an energy source in some stages of this complex process. A simplified summary equation for photosynthesis is:

$$\text{water} + \text{carbon dioxide} \xrightarrow[\substack{\text{chlorophyll}\\ \text{enzymes}}]{\text{energy}} \text{glucose} + \text{water} + \text{oxygen}$$

The glucose formed may be converted in the cell into other compounds such as fats, proteins, or starch; or it may be used as an energy source during cellular respiration. The use of oxygen-18 ("heavy oxygen") by scientists has shown that the oxygen involved in this process comes from the water. Carbon-14 has been successfully used to trace the pathways of carbon fixation in green plants. Photosynthesis is the source of most of the oxygen in the air. The rate of photosynthesis is affected by a number of factors including temperature, carbon dioxide concentration, wavelength of light, and availability of certain minerals.

(2) Chemical Aspects: Two sets of reactions occur in photosynthesis.

—Photochemical reactions (light reaction). During the light reactions the light energy absorbed by chlorophyll is used to "split" water molecules, releasing hydrogen atoms and oxygen gas.

—Carbon fixation reactions (dark reaction). In the second group of reactions, carbon dioxide molecules and hydrogen atoms participate in a series of chemical changes. These result in the formation of PGAL (phosphoglyceraldehyde) from which other compounds, including glucose, are synthesized. PGAL may be used in several ways: some is recycled to continue carbon fixation; some may be oxidized to provide energy for synthesis; and some may be used in the synthesis of carbohydrates, proteins, or fats.

b. Site: The photosynthetic process usually takes place within the chloroplast of a plant cell. In some algae, chlorophyll is not contained within a chloroplast. The chloroplast is a complex, highly organized plastid which contains the photosynthetic pigments of chlorophyll. Several types of chlorophyll are found in leaves; of these, chlorophyll *a* and chlorophyll *b* are the most important. The chlorophyll traps certain wavelengths of light energy (usually sunlight) and converts this energy into chemical bond energy inside the cell. Red light and blue light are highly effective. Chlorophyll reflects green light. Higher multicellular green plants have specialized organs; they reflect green light. Higher multicellular green plants have specialized organs, the leaves of which are the principal sites of photosynthesis. Two adaptations of leaves for food making are the comparatively large surfaces, which enable reception of light energy, and openings in the leaf, known as stomates, which permit the diffusion of carbon dioxide, water vapor, and oxygen between the leaf and the atmosphere. The size of the stomates is controlled by the guard cells. The stomates open into air spaces that are in contact with an internal moist surface which aids the diffusion of carbon dioxide from the air into the cells. Water for photosynthesis enters complex plants through the root hairs by osmosis and is conducted to the photosynthetic tissues through the vascular tissue.

2. Chemosynthesis: Some bacteria are able to synthesize organic compounds from inorganic sources without a need for light energy. These bacteria oxidize compounds of nitrogen, sulfur, or iron, and are able to use the small quantity of energy released by the oxidation. This energy is then used to manufacture food. An example is the nitrifying bacteria, which oxidize ammonia and nitrites, and play an important role in the nitrogen cycle.

B. *Heterotrophic Nutrition*: Heterotrophic plants, such as fungi and most bacteria, lack chlorophyll and are unable to synthesize their own complex organic compounds from inorganic compounds. These organisms obtain preformed organic molecules from other organisms by extracellular digestion and absorption. This method of energy capture is essentially the same as in animals.

II. DIGESTION

Plants lack a specialized digestive system, but the kinds of changes that occur in digestion are similar in all plants and animals. Plants store starches, lipids, or proteins in cells. These storage products are converted into the simpler usable forms as a result of enzymatic hydrolysis within the cells (intracellular digestion). The principal carbohydrate reserve of higher plants is starch. It is stored in large amounts in seeds, stems, and roots. Once digestion is completed, the end products may be used in the cell, or they may be transported to other tissues for use there. Heterotrophic plants secrete enzymes which hydrolyze foods outside the plant, changing them into simple organic compounds which can be absorbed and used for energy or converted into more complex compounds. Insectivorous plants carry on extracellular digestion of insects.

III. TRANSPORT

The process of material transport is similar in plants and animals. Higher plants have specialized absorbing structures and transport (vascular) tissues to provide cells with required materials. Like animals, complex plants have a transport system. Circulation in plants is referred to as translocation.

A. *Roots*: Roots and root hairs are structures specialized for anchorage and absorption.

1. Root hairs: Root hairs are elongated epidermal cells which increase the surface area of the root for the absorption of water and minerals. The movement of minerals and water from the soil into the plant is accomplished by active transport and diffusion processes.

2. Xylem: Xylem and phloem are plant tissues specialized for transport. These tissues extend from the roots to the edges of the leaves. The principal function of the xylem tissue is to conduct water and minerals upward in the plant.

3. Phloem: Phloem conducts organic food materials both upward and downward to plant tissues for immediate use or storage.

B. *Stems*: Although the anatomy of the stem differs from that of the root, the conducting tissues of the stem (xylem and phloem) are continuous with those of the root. Horizontal transport is also necessary in plants of large diameter, such as trees, if all cells are to receive nutrients. This is accomplished in vascular and pith rays by diffusion and active transport. The physical factors concerned with the upward transport in the xylem are transpiration pull, capillary action, and root pressure.

C. *Leaves*: Leaves contain veins which are extensions of the xylem and phloem of the stem.

IV. RESPIRATION

In green plants, respiration occurs 24 hours a day as in animals. Photosynthesis occurs only when light energy is present. The energy from solar radiation, captured by photosynthesis in the chemical bonds of organic molecules, is converted into ATP molecules during the process of respiration. Some ATP is formed during photosynthesis, but this process is not considered as a net source of ATP production.

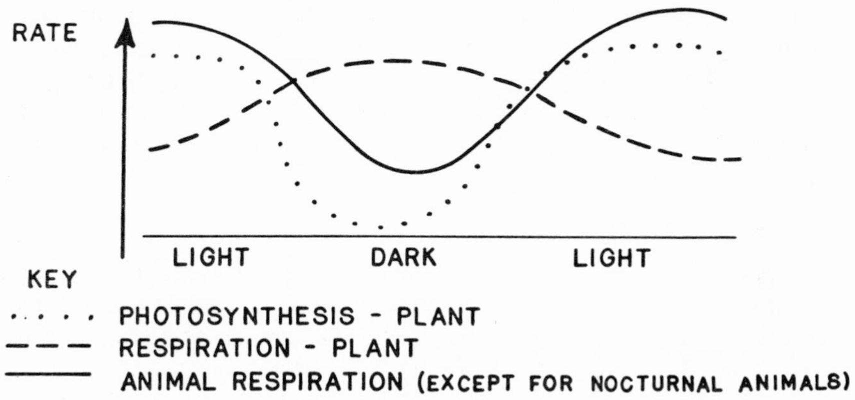

Diagram 30.

A. *Types*

1. Aerobic: Aerobic respiration occurs in the presence of free (molecular) oxygen. Oxygen and carbon dioxide move from areas of greater concentration to areas of lesser concentration by diffusion. Gases may enter or leave the large internal intercellular air spaces of the plant through the stomates and the lenticels of woody stems. Roots will also function as respiratory organs if the soil is well aerated. As the free oxygen content increases in a cell during photosynthesis, the gas will diffuse into the air spaces and may escape into the atmosphere or diffuse into cells having an inadequate supply. When photosynthesis is not occurring in the cells, oxygen may diffuse from the air spaces into the cells and from the atmosphere into the air spaces. As in animals, the mitochondria are the sites of the chemical activities associated with energy conversion in respiration.

a. Description: Many unicellular and multicellular plants release energy by aerobic respiration in a manner which is essentially the same as that in animals.

2. Anaerobic: Anaerobic respiration does not require free oxygen.

a. Description: Since molecular oxygen is not utilized in the release of energy from organic compounds in anaerobic respiration, it takes place whether or not oxygen is present. In the continued absence of oxygen, ethyl alcohol or lactic acid may be produced by a process called fermentation.

b. Chemical Aspects: Anaerobic respiration is similar in plants and animals.

(1) Alcoholic fermentation: This process begins with the formation of pyruvic acid with the same sequence of enzyme-catalyzed steps that produced it in animal tissues. If molecular oxygen is lacking, some simple plants, such as certain bacteria and yeasts, carry on alcoholic fermentation, producing ethyl alcohol and carbon dioxide as products of the reaction. The organisms derive their energy from the glucose bonds. A simplified summary equation for the alcoholic fermentation of glucose is:

$$\text{glucose} \rightarrow \text{pyruvic acid} \rightarrow \text{ethyl alcohol} + \text{carbon dioxide} + \text{energy}$$

Most of the released energy is stored in ATP bonds. However, most of the potential energy of glucose still remains in the bonds of alcohol. Alcohol fermentation has value in such industries as baking, brewing, and wine-making.

(2) Lactic Acid Fermentation: Lactic acid bacteria and certain molds can carry on lactic acid fermentation. In this process the bacteria or molds also obtain their energy from the bonds in glucose, and lactic acid is the product of the reaction. Lactic acid fermentation has commercial value in the food, fermentation, pharmaceutical, and chemical industries.

B. *Energy Yield*: Since most of the C-C and C-H bonds of the compound remain intact at the end of both alcoholic and lactic acid fermentations, only a relatively small amount of energy is released in comparison to the amount released in aerobic respiration.

Anaerobic respiration provides enough energy to support lower forms of life, such as bacteria and yeasts, whereas aerobic respiration provides the energy needed by higher plants. Yeasts may also respire aerobically.

V. EXCRETION

Plants have no specialized system for excretion comparable to that of higher animals. Since green plants are able to use many of the products of metabolism in synthesis, they give off only those products which are no longer useful to them. The main product of carbohydrate metabolism, carbon dioxide, is nontoxic to plants and can be re-used in the photosynthetic process. Some of the oxygen produced in photosynthesis is used by the plant for respiration. Excess carbon dioxide leaves the plant by diffusion through the stomates, lenticels and roots. Excess water moves as a vapor through the stomates and lenticels by the process of transportation. Since most of this transported water is not a metabolic waste, it is not a true excretory product. Some products, such as organic acids, which might be toxic, are stored in vacuoles where they are "sealed off" and cause no injury to the plant. Ordinarily, plants can use the ammonia and other nitrogen compounds derived from breakdown processes. These, along with additional absorbed nitrogen compounds, are used to synthesize amino acids and other compounds.

VI. SYNTHESIS

A. *Process*: Synthesis in a plant cell involves the formation of complex, functional compounds necessary for the maintenance of that cell, or in multicellular plants for the survival of other cells in the plant.

B. *Products*:

1. Structural: Both the production of new structures and the repair of existing structures in plant cells result from synthetic activities which form carbohydrates, fats, and protein. The glucose molecule synthesized as a result of photosynthesis can, in turn, be used for the synthesis—by dehydration—if more complex carbohydrates such as starch.

2. Other: A wide variety of chemical substances are synthesized by plants. These include: poisons, waxes, fibers, drugs, and flavorings.

C. *Outcome*: In addition to growth and repair in the cell, synthetic products may also serve to resupply specific compounds required by individual cells or by the entire plant (e.g., starch). The most abundant organic reserve is starch. Starch is insoluble or nearly so, and thus does not tend to affect the osmotic pressure in a cell. Auxins regulate plant growth, while scents and poisons are associated with reproduction and protection.

VII. REGULATION

Complex plants lack a nervous system but have chemical regulators, the best known (and the most common) of which are the auxins. The most widely distributed auxin is known as IAA—indolacetic acid. Gibberellins represent another class of growth-regulating compounds.

A. *Production of Auxins*: Auxins are a group of substances produced in the tips and roots, and subsequently transported to other parts of the plant. These function as plant growth regulators. Auxins have been found to stimulate rapid growth in young stems and to induce the development of flowering in a number of plants.

B. *Effects of Auxins*: Auxins are distributed through stems and roots, maintaining normal growth. Unequal distribution of auxins is brought about by stimuli such as light and gravity. Difference in auxin distribution brings about growth responses called tropisms. These responses are effected by accelerating or retarding the lengthwise growth of individual cells. Phototropism and geotropism are plant growth responses which may be explained in terms of unequal distribution of auxins. Auxins also appear to have some effect on cell division, and, as such, have commercial value as stimulators of rapid and extensive root formation on stem cuttings. Auxins also play a role in leaf and fruit abscission.

VIII. GROWTH

Growth in complex plants occurs in specific regions, whereas growth in animals may take place throughout the organism. Growth regions of complex plants are found in the root tips, tips of stems, and between xylem and phloem. These growth regions contain undifferentiated cells which are undergoing active cell reproduction. The growth tissue between the xylem and phloem (cambium) is responsible for growth in width while the growth tissues at the tips are concerned with growth in length.

ALGAE

AIM: To observe and study the various types of algae.

Algae are thallophytes. They are simple plants, often unicellular, which can range in size from those seen best under a microscope to the brown seaweeds which grow to remarkable lengths. One can distinguish the various phyla of algae through the presence of pigments that give the plants a characteristic color. Although all algae contain chlorophyll, many of the algae are blue, red, or brown because the other pigments mask the chlorophyll.

Algae can be found in freshwater ponds and streams as well as in the sea. They have no stems, roots, or leaves, and are nearly always associated with water.

MATERIALS NEEDED: Samples of the Blue-green algae such as Nostoc or Oscillatoria; Green Algae such as the Volvox or Spirogyra; Marine algae (red and brown) such as the Fucus and the Chondrus crispus; glass slides, cover glasses, dissecting needles, methyl cellulose, methylene blue, Lugol's solution, microscope.

PROCEDURE:

(A) Phylum Cyanophyta: Blue-green Algae

They have no chloroplasts but yet chlorophyll is present. Nuclear material is scattered throughout the cell. Many exist as a single cell while others form threads, or filaments.

1. Examine a gelatinous clump of Nostoc filaments. With a dissecting needle tease away a small amount of the plant and make up a wet mount.

2. Examine under low power of the microscope. You will see that the individual cells are bead-shaped and as you move the slide about the filament will look like a string of beads. The filaments are embedded in a jelly-like matrix. You will also see cells that appear empty. These cells are called

heterocysts and are the points where the filament may break. Make a drawing.

3. Place a small amount of Oscillatoria in a drop of water on a slide. Examine under low power. Watch the tip of a filament and you will see that the filaments gradually sway from side to side which gives the plant its name.

4. Examine Oscillatoria under high power. You will see small dots in the protoplasm. These are grains of starch. You will also notice that every now and then there will be cells that bulge on either side so that they look like a concave lens. These are called concave cells and the Oscillatoria filaments often break at this point. The rest of the cells are almost rectangular and a filament will be of uniform width since all the cells are almost of equal size.

(B) Phylum Chlorophyta: Green Algae

These are the most widely distributed plants and found wherever there is moisture. They vary from single-celled to many-celled branching filaments.

Protococcus is one of the most common of the unicellular forms and found as a green film on damp walls, rocks, and the bark of trees.

The Volvox live in colonies of hundreds of individual cells arranged on the surface of a hollow sphere surrounding a jellylike center. Each cell bears 2 flagella. The beating of the flagella causes the sphere to rotate. They reproduce by means of fission of a single cell or when specialized egg and sperm cells unite.

(1) Make a wet mount of a piece of Volvox and examine under low power. You may slow down the rotation by means of a drop of methyl cellulose.

(2) Change to high power and add a drop of methylene blue under the cover glass so that you can see the details more clearly. Draw and label flagella, cells, daughter colonies.

Spirogyra is a very common pond plant that is easy to identify because each cell contains a chloroplast in the shape of a helix or spring and starch is stored in pyrenoids along the chloroplast. Spirogyra reproduce by means of conjugation when the contents of one filament move across a conjugation tube into a cell of another filament. The result is a thick-walled zygospore which can withstand unfavorable conditions.

(1) Place a few filaments of Spirogyra on a slide and make a wet mount. Tease the filaments apart, add a cover glass and examine under low power.

(2) To show that the pyrenoids contain starch, add a drop of Lugol's solution to the slide and the starch granules will stain purple to black.

(3) Make a drawing of Spirogyra and label cell wall, chloroplast, pyrenoid, nucleus.

(C) Phylum Phaeophyta: Brown Algae

Most of the brown algae are marine forms and are similar to green algae in body type and methods of reproduction. However, the brown algae tend to have a more complex type of vegetative body as well as multicellular sex organs. In fact, they resemble vascular plants in organization but lack xylem, a cuticle, and special coverings for their sex organs.

The most common example of brown algae is Fucus or rockweed. It is found growing on rocks at the seashore. The male and female gametes are produced in separate cavities called conceptacles which are found in swollen areas, the receptacles, at the end of the branching thallus.

(1) Examine the entire plant. You will see a rootlike structure called the holdfast, the receptacles, and the air bladders.

(2) From the holdfast, which is rootlike in function, there is a stipe which in turn holds the frond. You will notice that along the margin of the frond, there are air bladders. These air bladders are so located that the plant can be held towards the surface when covered with water; they float the plant in the water.

(D) The Red algae:

Red algae have characteristic female reproductive structures called the carpogonium. The reproductive cells do not have flagella. They resemble the blue-green algae. They contain a red pigment but they are often green, brown, or purplish-black depending upon the pigments present and the amount of each. They are usually found in warm seas.

An example of red algae is Chondrus crispus.

(1) Examine a specimen. You will find it to contain a holdfast, a stipe and a frond.

CONCLUSIONS:

Algae are a very large and varied group and it is quite difficult to identify some of the more complex forms as algae because of their size and organs.

ELODEA

Elodea is a multicellular green plant which grows at the bottom of lakes and ponds. Because its leaves are only one or two cell layers in thickness they can readily be studied under the compound microscope. You can observe the structure of a "typical" plant cell which contains chlorophyll. Note that the previous plant cells you studied—those of the onion skin—did not contain this green substance.

PART I

PROCEDURE:

1. Remove a young leaf from the tip of an Elodea plant. (The leaves at the tip are not so heavy as those farther down on the stem, and are more

suitable for observation under the microscope.) Prepare a wet mount using tap water and a coverslip.

2. Use low power to scan the entire surface of the leaf. Note the spine cells at the margin of the leaf.

3. Make a low power diagram of a group of cells. Indicate and label the nucleus, cytoplasm, cell wall, and chloroplasts.

4. Measurement.

a. How many cell lengths occupy the diameter of the field under low power?

b. How many cell widths?

c. In a previous exercise you determined the size of the diameter of your low power field. This was . . . . . . μ.

d. Perform the following calculations:

(1) The approximate length of a cell . . . . . .

(2) The approximate width of a cell . . . . . .

(3) The approximate area of a cell . . . . . .

(4) How many cells are in one square millimeter of leaf surface?

5. Switch very carefully to the high power on the microscope. Focus up and down with the fine adjustment. How many layers of cells are there in the section which you are observing? . . . . . .

6. *Cyclosis*: If the plant is a healthy specimen, growing in warm, well-lighted water, you may see the chloroplasts moving in a circular pattern around the cell. Search the various portions of your leaf to find this phenomenon. If you do not find it, try warming the slide on the palm of your hand for one minute, or pass the slide quickly over a flame. (*Caution*: Do not heat the slide since heat will kill the cell!) This movement is caused by the movement of the cytoplasm and is known as cyclosis.

7. Make a high power drawing of two cells, labeling all parts.

PART II

8. Effect of salt water: Change the environment of your leaf to that of a highly concentrated solution of salt. Do this by placing a drop of salt solution at the side of the cover slip. Then draw this solution under the cover slip by using a small piece of paper toweling at the opposite side of the cover slip. Observe the cells again under low power. Can you now observe any difference in the cell membrane? (Formerly this was pressed tightly against the cell wall.) The shrinkage of the cell contents when placed in salt water is known as *plasmolysis*.

9. Explain the phenomenon of plasmolysis.

10. The opposite of plasmolysis is *plasmoptysis*. How could you cause plasmoptysis? Explain why it happens.

MINUTE STRUCTURE OF THE ROOT, STEM, AND LEAF*

AIM: To study the minute structure of the root, stem, and leaf and compare the structure of a very small grass seedling root with the structure of a large root such as a carrot.

MATERIALS: Grass seedlings, several days old, grown on plastic sponges; microscope slides and cover slips; medicine droppers; beakers of water; compound and dissecting microscopes; scalpels; tweezers; carrots; prepared slides of stem and leaf.

PROCEDURE:

A. Observe the plastic sponges on which grass seeds have been germinated. The young grass plants (seedlings) are 3, 4, and 5 days old. Observe the differences in the growth patterns.

B. Carefully remove one or two seedlings with the tweezers and place the seedlings on a microscope slide. Add 1 or 2 drops of water.

C. Examine the young plant under the dissecting microscope.

1. Make a sketch of one plant showing young blade and the root. Include all details that are visible. Identify root cap, root hairs, branch roots, if any.

D. Carefully cover the seedlings on the slide with the cover slip. Avoid forming air bubbles by adding more water at the edge of the cover slip. DO NOT PRESS down on the cover slip as this will crush the delicate young plant.

E. Under LOW POWER of the compound microscope examine the root of the grass seedling. Look at the tip of the root for a loose structure fitting over it like a thimble. This is the ROOT CAP.

2. What is the appearance of the cells of the root cap?

3. What is the function of the root cap?

F. Just behind the tip of the root you will find the root hairs.

4. Do they originate from epidermal cells or from cells deeper in the root?

5. How many cells make up a root hair?

6. What is the advantage to the plant of the thin walls of the root hairs?

G. Observe one root hair under HIGH POWER of the compound microscope.

7. Draw one root hair and show as much structure as you can observe.

8. Do you see any movement? Describe what you see.

H. Switch back to low-power magnification and observe the root above the region of the root hairs. This is the main body of the root. Find the

* Starred experiments throughout the book should be checked for materials at least two days in advance. Some materials must be prepared ahead of time.

central cylinder.

9. Can you see the cells of the cylinder? These will be differentiated as phloem and xylem cells.

10. Draw one or two cells from the central cylinder.

I. Outside the central cylinder are the cells of the cortex.

11. Draw one cortical cell.

12. What is the function of the cells of the cortex?

13. Are there branch roots?

14. How are they attached to the main root?

J. Observe the cross section and longitudinal section of a carrot.

15. What structures found in the grass seedling root can also be seen in the carrot?

16. What structures are not visible?

17. Summarize the functions of the root and the adaptations in structure which enable the root to perform its function.

K. Study prepared slides of the following and make drawings.

(1) Cross section of a leaf.
(2) Lower epidermis of a leaf.
(3) Cross section of a woody stem.
(4) Cross section of an herbaceous stem.

BREAD MOLD AND YEAST*

AIM: To study some examples of non-green plants.

INTRODUCTION: Bread mold and yeast are classified as plants but since they are not green, they cannot manufacture their own food. Molds are *fungi* and seem to grow everywhere. The molds are especially noticeable since they may spoil fruits, vegetables, baked goods, leather and even plants where they cause disease. Molds grow rapidly on many warm moist foodstuffs. If you have bread at home that does not have a mold inhibitor, you will find that there will be a growth of mold within a few days if you moisten the bread and place it in a warm spot.

Yeasts are also fungus plants that obtain energy for metabolism and reproduction by fermenting sugars into carbon dioxide and alcohol. This process is made use of in the raising of dough when the yeast is kneaded into the dough and set aside in a warm place. The dough will rise and form small air spaces. This makes bread light.

MATERIALS:

For each team of students: a culture of Rhizopus nigricans, two dissecting needles, forceps, glycerine and a dropper, slides and cover slips, culture of yeast in dilute molasses, silicone gum.

At demonstration table: Demonstration of bread mold under a binocular microscope; demonstration of bread mold contained zygospores under a compound microscope; fermentation tube with yeast.

PROCEDURE:

1. Take up a *minute* portion of the mold using the dissecting needles and make a wet mount using glycerine. (Avoid breathing in any of the bread mold spores.)

2. Use the low power of a compound microscope and scan the field. You will find that the natural position of the parts will be distorted. Find interesting areas and switch to the high power. Make separate drawings of each of the following: (a) rhizoids (b) stolons (c) sporangium which is probably broken. Label all parts.

3. At the demonstration table, the bread mold should be mounted in a silicone gum cell so that it will be clearly seen when under a binocular microscope.

4. Now make a slide of the yeast culture in a silicone gum cell by placing a drop of the culture on the slide. Observe under the microscope. Look for budding cells. Make a drawing.

5. Observe the yeast growing in a fermentation tube. You will notice the air space in the arm growing larger as time passes. The gas can be tested for carbon dioxide.

OBSERVATIONS AND CONCLUSIONS:

1. Write up what you have seen in the laboratory and include drawings.
2. What is the scientific name of the common bread mold?
3. Where does digestion of the bread occur?
4. What are the functions of the rhizoids?
5. What is the function of the sporangium?
6. Into what phylum and class do we classify bread mold?
7. What is the difference between sexual and asexual reproduction?
8. The demonstration microscope showed *zygospores* produced by a union of equal-sized cells from two hyphae in the process of *conjugation*. Is this asexual or sexual reproduction?
9. Where are the asexual spores produced?
10. What name is given to the division of a yeast cell into two unequal cells?
11. The fermentation of yeast may be represented by the following equation:

$$\text{sugar} \xrightarrow{\text{zymase}} CO_2 + \text{alcohol}$$

a. What is zymase?

b. What industry depends upon the carbon dioxide produced in this reaction?

c. What industry depends upon the alcohol produced?
d. What contribution did Pasteur make to this subject?

12. For what purpose do bakers use sodium propionate?
13. Give the scientific name of a mold which makes an antibiotic?
14. How is Roquefort cheese made?
15. Name two diseases caused by the growth of parasitic molds upon the skin.
16. A certain mold has been used extensively by G. W. Beadle in elucidating the mechanism of heredity. What is the name of this mold?
17. Name three physical conditions which favor the growth of molds.
18. What is mildew?
19. Why should towels be dried before being placed in a laundry hamper?

CARBON DIOXIDE IN PHOTOSYNTHESIS*

AIM: To see that carbon dioxide is an important chemical in the process of photosynthesis.

Photosynthesis is the process by which green plants change carbon dioxide and water into glucose.

MATERIALS NEEDED: Two wide-mouthed gallon jars with cover, a small beaker, concentrated sodium hydroxide solution, 2 seedlings of any type (both alike) in paper cups, Lugol's solution, .04% sodium hydroxide solution, 50 ml water charged with carbon dioxide or seltzer, burette, burette clamp, iron stand, 250 ml beaker, phenolphthalein, Elodea, water, 90 ml distilled or demineralized water, glass rod, 2 medium-sized beakers, graduated cylinder.

PROCEDURE:

(A) Carbon Dioxide and Photosynthesis

1. Place an Elodea plant in a beaker and cover with exactly 200 ml of carbon dioxide charged water.
2. In an identical beaker, place exactly the same amount of charged water without the plant.
3. Place both beakers side by side in a sunny place and allow them to remain there for at least one hour. More time will give better results.
4. Remove the plant and add 4 drops of phenolphthalein solution to both beakers.
5. Now follow the exact method used in part B of this exercise to calculate the number of grams of carbon dioxide in the water that contained the plant and the one that did not.

6. There should be more grams of carbon dioxide in the beaker that did not contain the plant.

(B) Measurement of Carbon Dioxide in Water

Carbon dioxide forms carbonic acid when it dissolves in water. Sodium hydroxide, a base, can neutralize the acid. The amount of the acid in solution can be determined by the amount of a base required to neutralize the acid.

In the following exercise, each milliliter of .04% solution of sodium hydroxide will neutralize .00044 grams of carbon dioxide in solution. We thus can determine the amount of carbon dioxide in solution, if we know how much sodium hydroxide solution has been used.

1. Pour 50 ml of the water charged with the carbon dioxide into a small beaker and add 3 drops of phenolphthalein.

2. Add 10 ml of the 0.4% sodium hydroxide solution to 90 ml of distilled water. Fill the burette with the resulting .04% sodium hydroxide solution.

3. Clamp the burette to the iron stand by means of the clamp. Record the burette reading.

4. Place the beaker containing the water and phenolphthalein under the burette opening. Drop by drop, add the sodium hydroxide solution. Stir the water after each drop with the glass rod. When the mixture remains pink, the carbonic acid has been neutralized.

5. Record the burette reading at this point and the amount of sodium hydroxide used is the difference between the readings at the beginning and at the end.

6. To calculate the number of grams of carbon dioxide in the water do the following:

Amount of sodium hydroxide used ×0.00044 gm per ml=the amount of carbon dioxide in the water

(C) The Need for Carbon Dioxide

1. Place a beaker of concentrated sodium hydroxide into a jar and cover it. Place next to the empty, lidded jar.

2. Twenty-four hours later, lower a seedling cup into each of the jars and close the jars immediately.

3. Place both jars in a sunny place for several days.

4. Take a leaf from each seedling and test for starch with Lugol's solution. The leaf from the jar without the sodium hydroxide will show a positive test for starch while the leaf from the seedling in the jar with the sodium hydroxide will show a negative test.

CONCLUSIONS AND DISCUSSION:

1. Why did the plant in the jar with the sodium hydroxide give a negative result when tested with the Lugol's solution for starch?

2. In Part B, what caused a change in color, from colorless to pink?

3. In Part A, how can you explain the larger amount of carbon dioxide in beaker #2 without the plant?

THE HILL REACTION*

AIM: To study the role of light reaction in photosynthesis as demonstrated by Robin Hill.

INTRODUCTION: The first notable contribution in the study of photosynthesis was made in 1905 by the English plant physiologist, Blackman. He demonstrated that photosynthesis is not a single photochemical reaction. It consists of two major reactions: the first is a rapid reaction and requires light energy for its acceleration while the second reaction does not use light energy and goes on equally well in either light or dark. However, it was not until 1937 that Robin Hill, an English biochemist, showed the possible nature of the light reaction.

Illumination of suspensions of chloroplasts in water in the presence of a suitable hydrogen acceptor (oxidant) results in the release of oxygen. This reaction also occurs in suspensions of grana from disintegrated chloroplasts. This phenomenon, now generally called the Hill reaction, can be represented by the following equation, A standing for a hydrogen acceptor:

$$2H_2O + 2A \xrightarrow[\text{chloroplasts}]{\text{light}} 2AH_2 + O_2$$

Some of the compounds which act as hydrogen acceptors in this reaction are ferrocyanides, chromates, certain quinones, and certain indophenols. By use of the 0–18 isotope of oxygen, it has been shown that the oxygen released in the Hill reaction, as in the over-all process of photosynthesis, comes from the water molecules.

The Hill reaction apparently consists in a photocatalyzed splitting of water molecules. A probable role of this reaction in the overall process is the formation of hydrogen atoms which are stored in the chloroplasts by combinations with some hydrogen acceptors. Subsequent chemical transformations involving hydrogen transfer utilize the hydrogen which has been formed and captured in this manner. Present evidence indicates that the hydrogen acceptor is triphosphopyridine nucleotide, TPN.

The primary role of light is the photolysis of water. The period during which CO_2 reduction can occur in the dark at the expense of a previous exposure to light is of very short duration.

Because the Hill reaction indicates that fragments of chloroplasts in the light decompose water to hydrogen and oxygen, it is of great interest in biology courses.

The overall procedure is to permit a suspension of chloroplasts from spinach to react in the light with a blue solution of indophenol (2, 6 dichlorophenol indophenol). As the indophenol takes up hydrogen it becomes reduced to a colorless form. By implication, oxygen is also released but no test for the oxygen is made in this procedure, nor are the bubbles of oxygen observed. The controls include the use of indophenol without chloroplasts and an additional pair of tubes in the dark.

PROCEDURE:

Since this is a rather complicated procedure, the following is for one demonstration but several teams can do the exercise if desired.

1. Prepare a solution of 0.01% indophenol by dissolving 0.1 gram indophenol in 1000 ml of demineralized water. Place this solution in a stoppered dark bottle and in the refrigerator until needed.

2. Prepare a cold solution of 10% sucrose by dissolving 10 grams of sucrose in 50 ml of demineralized water and add ice cubes to bring to the 100 ml mark.

3. Place 25 grams of fresh spinach leaves in a Waring blendor and add 100 ml of the cold sucrose solution (including ice cubes). Homogenize for 30 seconds at high speed.

4. Filter the homogenate through two layers of cheesecloth placed over a beaker.

5. Centrifuge the filtrate for five minutes at high speed. Meanwhile prepare another 100 ml of cold 10% sucrose as described in step 2.

6. While step 5 is in progress prepare four test tubes of cold indophenol solution, as follows: To 10 ml of ice water in each tube add 2 ml of the previously prepared indophenol solution. (The indophenol could be diluted to start with but if this is being done as a laboratory exercise, it gives the student practice in using pipettes and making dilutions.)

7. After the chloroplast suspension is centrifuged, pour off the supernatant fluid and re-suspend the chloroplasts in 2 ml of cold sucrose solution. This is now the relatively cell-free chloroplast solution which is used in the main portion of the exercise.

8. Now mark the tubes as shown in Diagram 31.

9. Add cold chloroplasts suspension to tubes 2 and 4: Using a medicine dropper add only one or two drops of the suspension prepared in step 7 and mix thoroughly the contents of each tube. A slight green tinge develops in these two tubes because of the chloroplasts. Wrap completely in aluminum foil the control tubes 3 and 4 whose contents are to receive no light.

10. Exposure to light: Place all four tubes into a beaker containing ice water and floating ice cubes. Expose to a 150 watt incandescent bulb placed a few inches from the beaker. Within 10 minutes tube 2 is decidedly more pale in color than tube 1. If the change is not apparent, add one additional drop of the chloroplast suspension to tube 2, and shake.

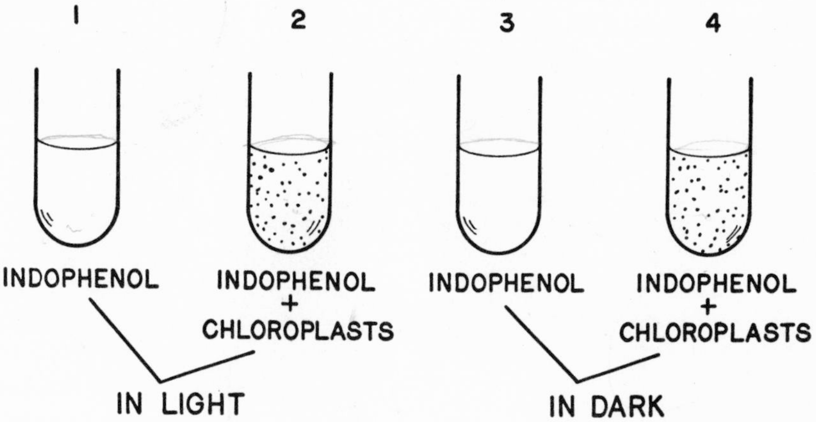

Diagram 31.

RESULTS:

Tube 2, containing the chloroplasts and exposed to the light, is a pale green. Tube 1 without the chloroplasts, is the same intensity of blue as tube 3. Tube 2 is much lighter in intensity than tube 4. Why?

CONCLUSION: Chloroplasts in the light produced a substance which reduced indophenol.

QUESTIONS:

1. Who was the first to really study photosynthesis?
2. How many reactions are involved in photosynthesis? Name them.
3. What causes the release of oxygen in photosynthesis?
4. What compounds act as hydrogen acceptors?
5. What is TPN?
6. What is the primary role of light?

CHROMATOGRAPHY AND FLUORESCENCE OF PLANT PIGMENTS*

AIMS:

(1) To separate several of the pigments of a plant.
(2) To observe the phenomenon of fluorescence of chlorophyll.

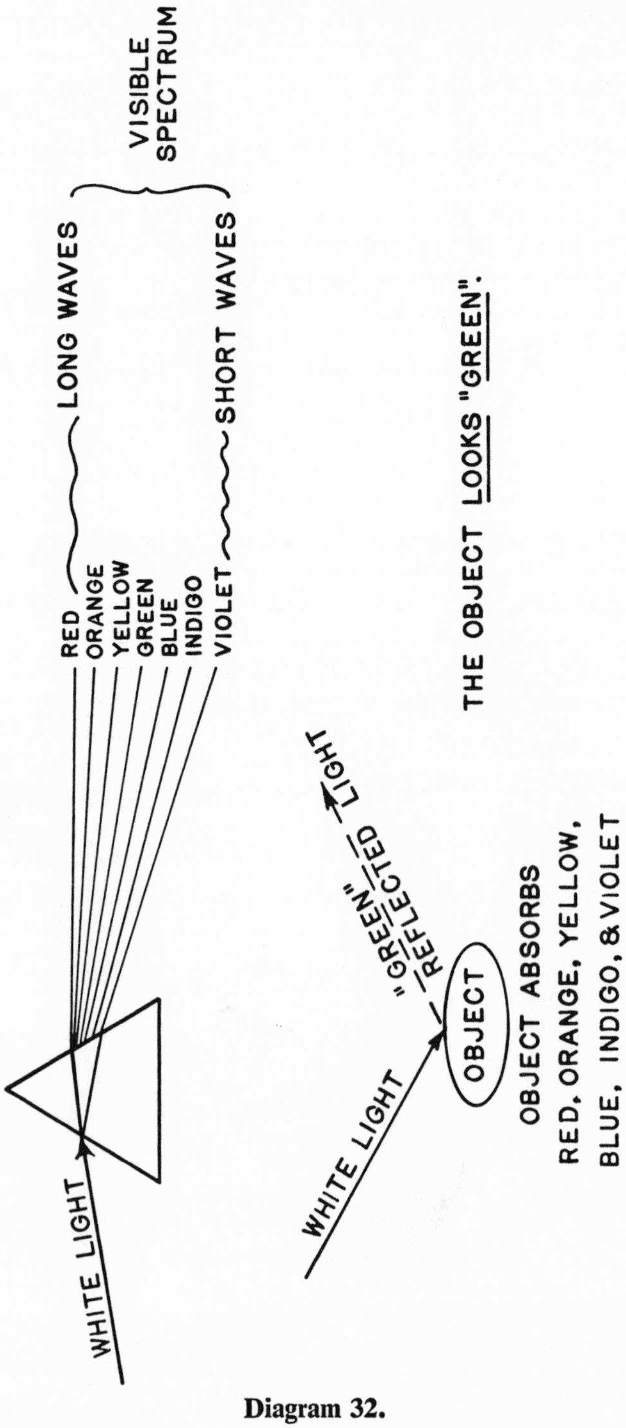
VISIBLE SPECTRUM
LONG WAVES
SHORT WAVES
RED
ORANGE
YELLOW
GREEN
BLUE
INDIGO
VIOLET
WHITE LIGHT
THE OBJECT LOOKS "GREEN".
"GREEN" REFLECTED LIGHT
WHITE LIGHT
OBJECT
OBJECT ABSORBS
RED, ORANGE, YELLOW,
BLUE, INDIGO, & VIOLET

Diagram 32.

INTRODUCTION:

Pigments: Among the pigments present are the following:

Chlorophylls: Chlorophyll *a*, chlorophyll *b*, chlorophyll *c*, chlorophyll *d*. These are various shades of green.

Carotenes: These may be white, yellow, orange, or red.

Xanthophyll: These may be bright yellow.

Anthocyanins: These water soluble pigments are present free in the cell cytoplasm instead of plastids. They may be red, blue, or purple.

In the autumn, when the chlorophyll in leaves disintegrates, the colors of the xanthophylls and anthocyanins become unmasked, causing the autumnal display.

Chromatography: We have discussed the principles involved in chromatography in general, and also in thin-layer chromatography in the experiments on paper and thin-layer chromatography in Unit I. Our reason for using thin-layer chromatography in this exercise is the speed with which the chlorophyll separates into its basic components.

Fluorescence: Diagram 32 reviews your understanding that white light is a mixture of various wavelengths. These may be dispersed by a prism into the component wavelengths, to form the visible spectrum. An object which looks green is one which absorbs all other wavelengths and *reflects* (or transmits) the wavelength which affects our eyes and brain with the sensation of green. When chlorophyll reflects green wavelengths, it is absorbing the other wavelengths of red, orange, yellow, blue, and violet. Since light is a form of energy, the wavelengths that are absorbed have the capacity to do work. Plants exposed only to green light cannot carry on photosynthesis, and soon die.

The structure of the chlorophyll *a* molecule is such that it can capture light energy. This seems to be by some form of resonance of the atoms

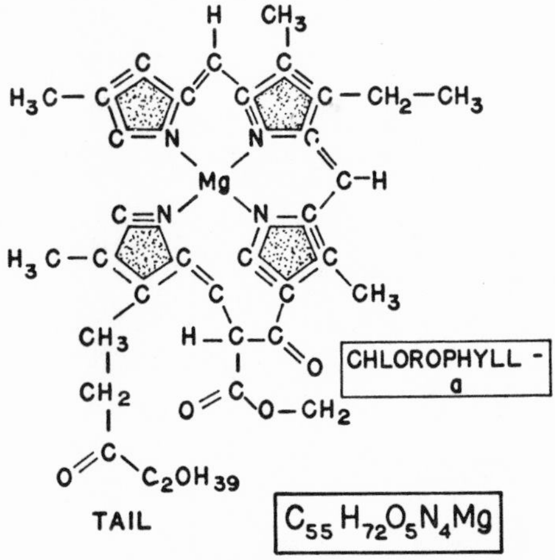

Diagram 33.

comprising the molecules. Examine the structure of chlorophyll molecule $C_{55}H_{72}O_5N_4Mg$ as shown in Diagram 33. Observe that it is composed of four complex carbon-hydrogen rings which are joined into a large ring, the "head." In the center of the head is a magnesium atom. It is interesting, and probably of evolutionary significance, to point out that there is another kind of molecule which resembles greatly the chlorophyll head except that there is an iron atom in place of magnesium. This is the *heme* molecule, the colored component of hemoglobin. The remainder of the chlorophyll molecule is a long "tail" of $C_{20}H_{39}$.

When red-orange and blue-violet light is absorbed by the atom of the chlorophyll molecule, the atom may become *excited*. That is, an electron may be displaced from its normal orbit farther away from the nucleus of the atom. (See Diagram 34.) This excited state lasts for a very short time—

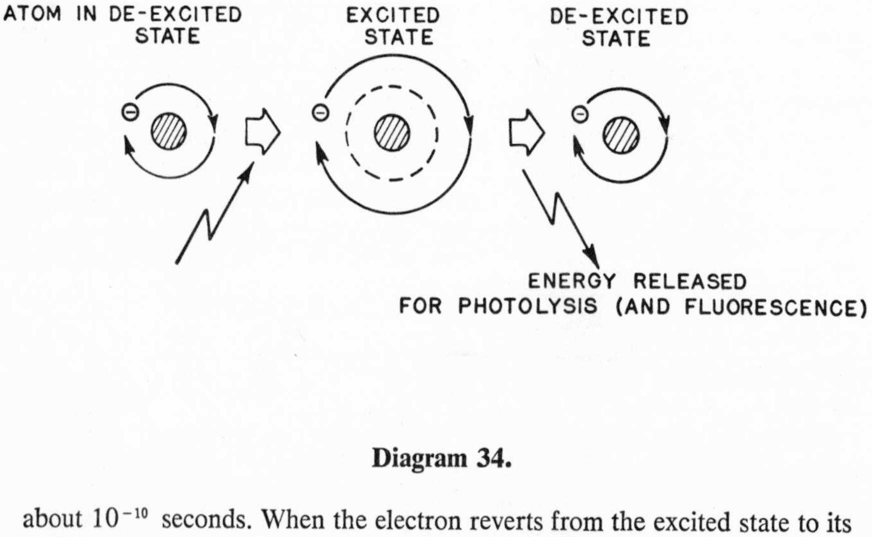

Diagram 34.

about 10^{-10} seconds. When the electron reverts from the excited state to its position, the atom releases the energy which it had absorbed. Most of it presumably is used in splitting water (see Diagram 35), but about 1% is emitted as a brief flash of light called *fluorescence*. Inasmuch as the light which is emitted has less energy than the light which was originally absorbed, it is of a longer wavelength and thus appears red.

PROCEDURE:

Preparing the Chloroplast Pigments

Materials Needed: Dried parsley leaves (10 grams); 25 ml acetone; 50 ml diethyl ether; mortar and pestle; funnel; beaker.

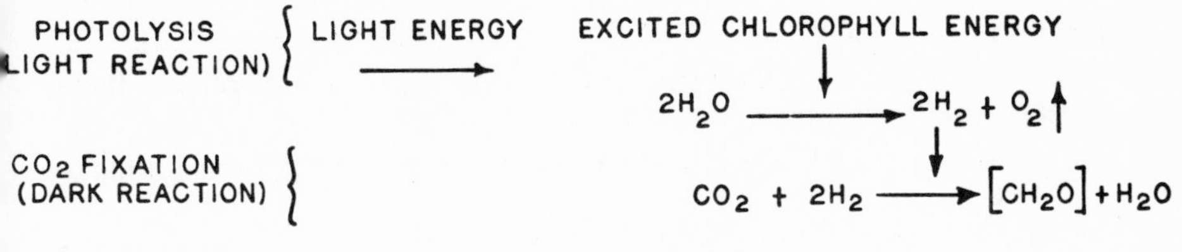

Diagram 35. General Pattern of the Two Phases of Photosynthesis.

The dried parsley leaves may be obtained from the spice shelf of any supermarket. The brand is not important. All give the same excellent results.

1. Weigh 10 grams of parsley and pulverize it to a fine powder with a mortar and pestle. Place the leaves in a beaker.
2. Add 25 ml acetone and 50 ml diethyl ether (sulfuric ether) to the powdered leaves. Mix thoroughly.
3. Allow the mixture to stand, covered, for at least 20 minutes, then mix again.
4. Filter the mixture. The filtrate is the chlorophyll extract that will be used for chromatography.
5. Store the extract in brown cork-stoppered bottles (rubber stoppers deteriorate) in the refrigerator. Colors are more vivid when the extract is left undisturbed for several days before use.

Preparing the Microscopic Slides (Microchromatic Plates)

Materials Needed: 21 grams Silica Gel G (Merck); 33.5 ml chloroform; 16.5 ml methanol; microscope slides; Coplin jar; stirrer.

Procedure: Method of preparation in experiment on Thin-layer Chromatography in Unit I. There will be a quantity sufficient for a glass of 20.

Spotting the Slides: Only minute amounts of the chloroplast solution are needed on the chromatoplate. Satisfactory spotting pipettes for school use may be made by drawing out melting-point tubes with the aid of a Bunsen burner.

1. Fill the pipette with the chloroplast solution by capillary action. The first drop may be too large and should be extruded onto a piece of scrap paper.
2. In placing a drop on the silica gel, try to touch only the liquid on the surface without breaking the surface. Ideally, each drop should be small enough to spread no more than 1 to 2 mm in diameter. Since the chloroplast pigments are dissolved in volatile acetone-ether, the drop dries immediately without the need for hot-air driers.

3. Arrange twenty drops adjacent to each other to form a line across the plate. The spotting should be about 1 cm from the bottom of the slide.

CAUTION: The room should be well ventilated at all times because of the extreme volatility of the liquids used.

Developing the Chromatogram:

Material: 2 volumes iso-octane; 1 volume acetone; 1 volume diethyl ether; Coplin jars; micropipettes; coated slides.

NOTE: Twenty-five (25) ml of this developing solvent should be sufficient for the Coplin jars for the entire class.

CAUTION: No flames in the room.

Coplin jars are convenient to use as developing containers, though the grooves should not be used for they will break the silica gel film. If these are unavailable, any small covered jars may be used. Aluminum foil or Saran Wrap may be fashioned into covers.

Procedure:

1. Into the chromatojar, introduce the developing solution to a depth of about 0.3 cm. For a Coplin jar, this requires a volume of about 2 ml. The height of the liquid in the jar should be below the spots on the microscope slide.

2. Cover the jar for a few moments so that the vapors fill the container.

3. Now insert the spotted microscope slide into the chromatojar and cover the jar.

4. The solvent rises, and within 2 minutes a brilliant array of green and yellow bands appears. Within 5 minutes the solvent front approaches the top of the silica gel layer and the slide should be removed for examination.

NOTE: The chromatogram is air-dried within 1 minute. No ovens or developing agents are needed. The pale yellow bands may begin to fade in light, but good observations can be made for several hours. Some slides are good for several days.

Analyses of Chloroplast Pigments

Identification of the Bands: For the average high school class, it is unnecessary to go beyond the observation that the green chloroplast solution has been separated into blue-green chlorophyll *a*, green chlorophyll *b*, yellow carotenes and yellow xanthophylls.

Rate of Flow: The rate of flow is the ratio of the distance traveled by the dissolved substances to the distance traveled by the solvent. To measure the R_f of the various components of chlorophyll, measure the distance the solute traveled from the original dot. Each band will have traveled a different distance. Then measure the distance the solvent traveled from the spots.

The R_f of any substance is constant for a given set of circumstances. When all of the distances have been measured, the R_f of each substance can be calculated individually by using the following formula:

$$\text{Rate of flow } (R_f) = \frac{\text{Distance traveled by the solute}}{\text{Distance traveled by the solvent}}$$

Absorption of light by Chlorophyll: The mixture of chloroplast pigments may be used for roughly indicating the wavelengths which are absorbed by chlorophyll. Shine a strong white light (as from a slide projector) through a triangular prism onto a screen so as to form a spectrum. Now interrupt the white light with a bottle of chloroplast pigments. The green portion of the spectrum becomes much more dim, indicating the absorption of the wavelength for green.

To demonstrate fluorescence, use the same mixture of chloroplast pigments. When exposed to a strong white light, the green fluid appears red when seen from certain directions. (The mixture should be placed in a square-sided bottle.)

OBSERVATIONS:

A report should be made on the experiment which should consist of the following:

A. Make an outline drawing of the microscope slide. Indicate the heights of the original spot, the spots obtained, and the height to which the solvent rose.

B. Calculate the R_f of the various spots, using the formula.

C. Conclusions: What conclusions can be reached as a result of doing this exercise on chloroplast pigments?

D. Answer the following questions:

1. Why do leaves turn yellow and brown in the fall?
2. What kind of pigment is responsible for the purple color of flowers?
3. Give two advantages of chromatography over the older chemical methods of analysis.
4. Explain the mechanism of chromatography.
5. What is an advantage of TLC over paper chromatography?
6. What names are given to the two major portions of the chlorophyll molecule?
7. Explain why a traffic light appears red.
8. What evidence is given in the introduction to indicate that plants and animals are related?

THIN-LAYER CHROMATOGRAPHY TO IDENTIFY PIGMENTS IN VARIOUS PLANT PARTS*

AIM: To show that chlorophyll can be present in other parts of a plant not usually associated with the process of photosynthesis.

INTRODUCTION: Students are taught that chlorophyll is present in leaves for the sole purpose of manufacturing food for the plant. However, there are other parts of a plant that have similar pigments. There are green seeds, flowers, and fruit, and the green color is of a similar nature to that of chlorophyll found in the leaves.

MATERIALS: Dried parsley leaves, dried green peppers, dried chive leaves, dried green onion leaves, dried split green peas for seeds, broccoli stems, broccoli flowerets, carrots, 200 ml acetone, 400 ml diethyl ether, 21 grams of Silica Gel G (Merck), 33.5 ml chloroform, 16.5 ml methanol, microscope slides, Coplin jars, capillary pipettes, graduates, iso-octane, acetone, diethyl ether, blendor.

The parsley, green pepper, chives, and green onions can be obtained, dried, from the shelves of most supermarkets. Split green peas can also be found in the supermarket. Since the carrot and broccoli stems and flowerets are not available dried, these can be purchased fresh and dried in a slow oven until all moisture evaporates.

To facilitate drying the fresh plant parts, the broccoli stems and flowerets should each be chopped very fine and the carrot should be grated. The small pieces are then spread separately on large shallow trays and placed in the oven at about 220°F for several hours.

CAUTION: Be certain that the materials do not dry too quickly and turn brown.

PREPARING THE CHLOROPLAST PIGMENTS: Extraction of the pigments should be done in exactly the same manner for each of the plant specimens. The method has been described in the preceding experiment.

PREPARING THE MICROSCOPIC SLIDES: Follow directions as outlined in the exercise on thin-layer chromatography in Unit I.

DIPPING THE SLIDES: Follow directions as outlined in the exercise on thin-layer chromatography in Unit I.

SPOTTING THE SLIDES: Follow directions as outlined in preceding experiment.

DEVELOPING THE CHROMATOGRAM: Follow directions as outlined in the preceding experiment.

FINDINGS AND CONCLUSION:

Superficial examination of the developed chromatograms reveals that every green extract contains chlorophylls, carotene, and xanthophylls, while the carrot contains only the carotene. To establish the uniformity of position for the bands in the various extracts, the height of each of the bands from the point of application to the level to which it has moved within the specified time can be measured.

As a result of computing the R_f values for each of the separated pigment bands, it was shown that all parts of the plant that are green in color contain xanthophyll, chlorophyll *a* and *b*, carotene, and phaeophytin; all contain chlorophyll *b* with lesser amounts of chlorophyll *a*. The carrot has just one band, the carotene, which is on the same level as the carotene band on the other seven slides.

Regardless of the location of the pigments in the plant, the green coloring contains the same pigments.

FLOWER PIGMENTS*

AIM: To see what happens to flower pigments when acids and bases are present.

Acids and bases are able to diffuse into the cells and change the color of many different plant pigments within the cell. They can do this since the ionic particles of acids and bases are small in size. However, the molecules of the pigments are too large and cannot diffuse out of the cell.

In this exercise we shall attempt to show that the pigments change color as a result of contact with an acid or base but upon neutralization, the normal pigmentation will return.

MATERIALS NEEDED: Microscope, fresh flowers of various colors, 0.1N HCl, 0.1N NaOH, 10% solution of ammonia, 10% solution of vinegar, glass slides, tap water, marking pencil, broccoli flowerets.

PROCEDURE:

1. Make a 0.1 N solution of NaOH by dissolving 4 grams of solid NaOH in 100 ml of water.
2. Make 0.1 N solution of HCl by mixing 1 ml of concentrated HCl in 100 ml of water.
3. Use as many slides as there are different colored flowers used. Mark one end of the slide, A for acid, and the other end B for base and W for water in the center.
4. Take one each of the different colored petals and tear the petals into 3 pieces.
5. Place one piece on the A, one piece on the B, and one piece on the W.
6. Add 2 drops of the acid to the petal marked A, 2 drops of the base to the petal marked B, and 2 drops of tap water to the one in the center marked W.
7. Observe the slide under the microscope to see any color changes within the cell. Record your results on a chart that you can make up as follows:

Sample	Natural color	Acid	Water	Base

8. Observe all the samples given you and then do the same using the vinegar and ammonia. Are there any differences in the pigment changes?

9. Take a fresh broccoli floweret and macerate it. Follow the same procedure outlined above. What happens?

10. Some pigments are water soluble. If you want to determine which of the pigments are water soluble, macerate the petal and boil in water. If there is no change in the color of the petal, it is not water soluble. Alcohol usually extracts those pigments not soluble in water. Try those flowers that did not respond to the water.

CONCLUSIONS AND DISCUSSION:

1. How did the acid and base reagents get into the cells to cause the color change?

2. Do all the flowers, regardless of color, change in the same way when acid or base is added?

3. Why was there no color change with water?

4. Are similar pigments found in some flowers as in other parts of the plant such as leaves or stems?

5. Can the pigments from flowers be used as dyes?

HOW DO PLANT CELLS MAKE STARCH FROM SUGAR?*

AIM: To learn whether starch found in plants has been synthesized from the sugar produced in the process of photosynthesis, or if this synthesis is independent of photosynthesis.

MATERIALS NEEDED: Paper carton, black construction paper, albino tobacco seeds, iodine solution, 8% glucose solution (by weight), flats for germinating seeds, test tubes, test tube racks, petri dishes, agar, glucose-l-phosphate, cheesecloth, sawdust, electric blendor, potato slices, beakers, funnel, filter paper, razor blades.

PROCEDURE:

1. Start the seeds germinating in sawdust.

2. Prepare your dark box by either lining a large paper carton with black construction paper, or covering the floor with plastic and draping flaps of paper over the open side which will fold over to cover the contents of the box. The box should be placed near a source of heat.

3. Prepare a medium containing 2% agar and 0.5% glucose-l-phosphate in water. Boil the medium, pour into petri dishes and allow to solidify.

4. Prepare the potato extract by putting potato slices into the blendor

and blending to a pulp. Strain through several thicknesses of cheesecloth into a beaker. Allow the beaker to stand for several hours so that the starch grains remaining will settle to the bottom of the beaker. Decant the filtrate carefully into another beaker so that the juice is starch-free. Test a few drops with iodine to make certain that this is so.

5. When the seedlings have reached the height of 2 inches or so, remove them from the sawdust and remove any of the kernels still remaining to eliminate any source of sugar.

6. Number the test tubes from 1 to 8 and do as follows to each one.
   (1) Green seedling in water and kept in the light.
   (2) Albino seedling in water and kept in the light.
   (3) Green seedling in water and kept in the dark box.
   (4) Albino seedling in water and kept in the dark box.
   (5) Green seedling in the glucose solution and kept in the light.
   (6) Albino seedling in sugar solution and kept in the light.
   (7) Green seedling in sugar solution and kept in the dark box.
   (8) Albino seedling in sugar solution and kept in the dark.

7. Keep the dark box and the test tube rack side-by-side in a warm place for 24 hours.

8. Meanwhile, to demonstrate enzyme action, place four drops of the potato extract, separately and well spaced, on the agar medium in the petri dish. The extract contains phosphorylase which is an enzyme that converts sugar to starch.

9. Test one drop immediately for starch, using the iodine solution. Test the remaining drops one at a time at 15-minute intervals. You will find that at first there is no starch and that the intensity of the blue-black color of the test will increase with time as more sugar is synthesized into starch.

10. At the end of 24 hours, take a leaf from a region close to the stem of each of the seedlings and test for starch with the iodine solution.

11. The results should be as follows in the test tubes:
   (1) starch
   (2) no starch
   (3) no starch
   (4) no starch
   (5) starch
   (6) no starch
   (7) starch
   (8) no starch

CONCLUSIONS AND DISCUSSION:

(1) We can thus conclude that green cells can make starch without sugar only in light.

(2) Albino plants regardless of light cannot make starch, thus showing that the chlorophyll of the green plants is necessary.

(3) Green or not, light or dark, plant cells can only make starch if sugar is available.

(4) Why did we use sugar solutions in this test?

THE EFFECT OF DAY LENGTH UPON FLOWERING*

AIM: To determine how plants respond to the day length in respect to their flowering.

The length of day is called photoperiod. The length of the day varies with the season. We, here in the Northern Hemisphere, are aware of the fact that the day length increases during the spring, reaching its maximum on June 21st. Then length of the day decreases until it reaches its minimum on December 21st.

The response of the plants to the length of the day is called photoperiodism. This response is of two types. They are the long-day and short-day responses. There are plants that will flower when the day is shorter. These plants are really long-night plants and are called S-D plants or short-day. Those plants that flower when the day is longer are really short-night plants and are called L-D plants. There also are plants to which the day length is not important; they flower in response to other conditions. These plants are called day-neutral plants.

In this exercise you will examine the effects of the various photoperiods on the flowering in plants.

You will use the radish for the long-day plants and the cocklebur for the short-day plants.

MATERIALS NEEDED: For each student of the team, there would be 2 trays of 10 seedlings each of the radish and cocklebur, very powerful artificial light, carton with cover and a dark cloth to cover the box, hand lens, marking pencil.

PROCEDURE:

1. First germinate the radish seeds and the cocklebur seeds by presoaking for 24 hours and then placing into the trays of soil, 2 trays for each type of seed. Water, and when the seedlings have reached a height of about 2 inches, remove those seedlings that do not seem as healthy as others. Leave just 10 seedlings of a type in a tray.

2. Now mark one tray of the radish S-D and the other L-D. Do the same for the cocklebur seedlings.

3. To simulate L-D (short-night) conditions, the trays marked L-D must be handled so that they get 8 hours of sunlight, 8 hours of artificial light (under the lamps), and 8 hours of darkness in the carton. Follow

these directions for a tray of radish seedlings and a tray of cocklebur seedlings.

4. To simulate S-D (long-night) conditions, the trays must be handled so that they get 8 hours of sunlight, at a window, and 16 hours of darkness in the carton.

5. Water the plants every day and examine the plants at least once a week. Are there any flower buds? Use a hand lens for examination if necessary.

6. Allow about 4 weeks for appearance of flowers.

7. If no flowers appear at that time, remove the buds from the plant and dissect carefully on a wet slide to determine whether the bud contains flowers or buds. Use a dissecting microscope.

You should find budding, at least, in the radish plants under L-D conditions and in the cocklebur plants under S-D conditions.

CONCLUSIONS AND DISCUSSION:

We thus see that the length of day is important for the flowering of a plant. This accounts for the variation in the time of year that different flowers bloom.

UNIT VI

REPRODUCTION AND DEVELOPMENT

I. CELL DIVISION

All cells arise from other cells by cell division. The process is complex and involves both nuclear and cytoplasmic division. In unicellular organisms, the consequences of cell division may be considered reproduction; in multicellular organisms, cell division generally accounts for growth in size. Cancer may be considered as a kind of abnormal cell division.

A. *Mitotic Cell Division*: Mitotic cell division usually involves two aspects:

MITOSIS—an orderly series of complex changes in the nucleus, normally resulting in a precise duplication of the chromosomes and in the distribution of a complete set of chromosomes to each daughter cell.

CYTOPLASMIC DIVISION—division of the cytoplasm which may occur during or after mitosis.

1. Processes

a. Mitosis: The process of mitosis involves:

—replication of each chromosome during the non-dividing period, resulting in double-stranded chromosomes. Individual strands are known as chromatids.

—disintegration of the nuclear membrane during the early stages of division

—synthesis of a spindle apparatus

—attachment of double-stranded chromosomes to spindle fibers by centromeres

—replication of centromeres, permitting separation of single-stranded chromosomes and their movement to opposite ends of the spindle

—nuclear membrane formation around each set of newly formed chromosomes, restoring the non-dividing state of the nucleus (although the chromosomes are single-stranded at this point)

The major result of mitosis is the formation of two daughter nuclei which are identical to each other and to the original nucleus in number and types of chromosomes.

b. Cytoplasmic division: Division of the cytoplasm usually, but not always, accompanies mitosis. In animal cells, division of the cytoplasm is accomplished by a furrowing of the cell, thus separating the two nuclei. In plant cells, a cell plate, on which cell walls form, is synthesized, thus separating the newly formed nuclei.

2. Comparison Between Plant and Animal Division: Mitosis is similar in plant and animal cells. In most seed plants, the spindle apparatus appears to be synthesized directly by the cytoplasm without centrioles. Centrioles have been detected in lower forms of plant life. The methods of cytoplasmic division differ in plant and animal cells.

B. *Meiotic Cell Division*: Meiosis is a special form of nuclear division which is associated with sexual reproduction and the formation of haploid nuclei:

1. Process: The process of meiosis:

a. First Meiotic Division

—replication of each chromosome occurs early in the first meiotic division

—synapsis (Intimate pairing) of homologous chromosomes. During synapsis, the chromosomes often twist about each other. Homologous chromosomes are chromosomes that pair during meiosis. In diploid cells destined to produce gametes, the even number of chromosomes exist as homologous pairs

—alignment of homologous pairs on the spindle

—disjunction of homologous pairs, one of each pair moving to opposite ends of the spindle

b. Second Meiotic Division

—alignments of double-stranded chromosomes on spindle

—replication of centromeres

—migration of single-stranded chromosomes toward opposite ends of spindle. The distribution of homologues (members of a pair of homologous chromosomes) between the two resultant nuclei is random.

As a result of meiosis, also known as reduction division, diploid (2n chromosome number) primary sex cells will become specialized reproductive cells with a haploid(n) nucleus.

2. Comparison with Mitotic Cell Division: Mitosis is associated with growth and asexual reproduction; meiosis, with sex cell production. As a

result of mitosis, daughter cells have the same chromosome complement as the original cells. As a result of meiosis, daughter cells have one-half the chromosome complement of the original cell.

II. REPRODUCTION

A. *Asexual*: Asexual reproduction is any method of producing new organisms which does not involve fusion of nuclei. The new organisms develop from undifferentiated cells of the parent, essentially by mitotic cell division.

1. Types: Many unicellular organisms reproduce asexually by fission or budding. Fission occurs with equal division of the cytoplasm. In budding, cytoplasmic division is unequal. Budding may also occur in some multicellular animals. There is evidence of mitotic cell division in most cases of fission or budding.

In many plants and some simple animals, single, specialized cells, known as spores, are released from the parent and can develop into new individuals.

Some multicellular plants reproduce asexually by vegetative propagation. The presence of undifferentiated tissue provides the source of cells from which new plants may develop from the parent by vegetative means. Examples of vegetative propagation include: cuttings, bulbs, stolons, tubers, and corms.

Generally, the invertebrate animals possess more undifferentiated cells than do the vertebrate animals. Thus, asexual reproduction is characteristic of many of the invertebrates.

2. Results: Since undifferentiated cells are produced as a result of mitosis within the parent, the offspring are genetically like the parent, unless mutations occur in the cells destined to become new individuals.

B. *Sexual*: The sexual reproductive cycle involves the production by adult organisms of function sex cells (gametes), the fusion of the nuclei of sex cells at fertilization to form a zygote, and the development of the zygote which may develop to the adult stage, completing the cycle.

1. Animals

a. Gamete Formation: Male and female gametes are usually developed in specialized organs called gonads. Males possess testes, which produce the male gametes, sperm; females possess ovaries, which produce the female gametes, eggs. Organisms which contain functioning male and female gonads are known as hermaphrodites. Hydra and earthworm are examples of hermaphrodites.

(1) Male: Four monoploid sperm cells usually develop from each primary sex cell that undergoes meiosis in the testis.

(2) Female: Only one monoploid egg cell is usually formed from each primary sex cell that undergoes meiosis in the ovary. The accessory cells, known as polar bodies, which result from meiotic divisions, may

degenerate. Meiotic divisions, in which the chromosome number is reduced from diploid to monoploid, are sometimes referred to as maturation divisions. At the beginning of the differentiation process of the egg, food (yolk) is deposited within the cytoplasm.

b. Zygote Formation: Fertilization is the union of the egg nucleus, containing the n chromosome number, with the sperm nucleus, also containing the n chromosome number, to form a zygote containing the 2n chromosome number. Thus, the species number of chromosomes is restored and homologous pairs of chromosomes are reformed.

Parthenogenesis, the development of an egg without fertilization, is a modification of sexual reproduction. Parthenogenesis may occur naturally as in the production of male bees. It has been achieved experimentally with the eggs of such animals as frogs, rabbits, and turkeys. Interpretations differ concerning parthenogenesis as a form of sexual reproduction, since it does not involve fertilization.

(1) External Fertilization: Reproduction in most vertebrate animals in water is characterized by external fertilization, since the gametes fuse outside the body of the female.

(2) Internal Fertilization, within the body of the female, is characteristic of animals that reproduce on land. Internal fertilization provides a moist environment for the transport of motile sperms.

The number of eggs produced is related to the following factors which affect the chances for survival:

(a) type of fertilization—external or internal
(b) type of early development—external or internal
(c) amount of parental care.

Generally, animals that carry on external fertilization produce more eggs than animals that carry on internal fertilization. This is an advantage to the species since the chances for survival of externally fertilized eggs are not as great as those for the survival of internally fertilized eggs.

c. Embryo Formation: The zygote, or fertilized egg, immediately begins to go through a series of mitotic divisions, known as cleavage, which produces a multicellular embryo. Cleavage is thought to be stimulated by fertilization. Factors initiating cleavage in cases of parthenogenesis are not fully understood.

2. Plants

a. Gamete formation: The processes of meiosis and fertilization occur in the flower, a structure specialized for reproduction in higher plants. The flower may contain the male reproductive organ, the stamen, and the female reproductive organ, the pistil. In some species, certain flowers contain only stamens (male), while other flowers contain only pistils (female).

(1) Male: The diploid cells of the anther, a part of the stamen, produce pollen grains which contain haploid nuclei.

(2) Female: The ovule developing within the ovary of the

pistil, contains the haploid egg nucleus.

b. Zygote Formation

(1) Pollination: The transfer of pollen grains from the anther of the stamen to the stigma of the pistil is known as pollination. Pollination may be accomplished by wind, water, insects, and birds. Flowering plants often have the problem of sperm transfer in a dry, external environment. This is partially solved by the presence of the thick wall of the pollen grain which prevents dehydration of its contents during its transfer to the female reproductive organ. The pollen grain germinates on the stigma and forms a tube which extends into the ovule. Sperm nuclei are formed at this time in the pollen tube.

(2) The union of male (n) and female (n) nuclei in the ovule results in a diploid (2n) nucleus which develops into a new plant.

c. Embryo Formation: Inside the ovule, the zygote undergoes a series of mitotic divisions which results in a multicellular embryo.

III. DEVELOPMENT

Development refers to the growth and change of the fertilized egg to the adult form. The process involves cleavage, differentiation, and growth. Regeneration, the process of replacing lost parts in an organism after damage or removal, may be considered a form of development, since it does involve cell division, differentiation, and growth.

A. Animals

1. Embryonic Development

a. Cleavage and differentiation: Cleavage leads to the formation of the blastula, a single layer of cells forming a hollow ball. One side of the blastula becomes invaginated (gastrulation) forming the gastrula, a two-layered structure, having an inner layer, the endoderm, and an outer layer, the ectoderm. As development continues, a third layer, the mesoderm, forms between the ectoderm and the endoderm. These three embryonic layers differentiate and give rise to the various tissues, organs, and systems of the multicellular animal.

b. Site of Embryonic Development

(1) External: External development occurs outside of the female's body.

(a) In water: The eggs of most fish and amphibians are fertilized externally and develop externally in a water environment. The developing embryo's source of food is the yolk stored in the egg.

(b) On land: Eggs of birds, many reptiles and a few mammals develop externally in a land environment after internal fertilization. The developing embryo's source of food is the yolk stored in the egg. Some adaptations for animals which develop externally on land are a shell which provides protection, and membranes which help provide a favorable environment for embryo development. These embryonic membranes include:

—the chorion which lines the shell and surrounds the embryo and other membranes. It serves as a moist membrane for gaseous exchange.

—the allantois which functions in respiration and excretion.

—the amnion which encompasses the amniotic fluid This fluid provides a water environment and protection from shock or adhesion to shell.

—the yolk sac which surrounds the yolk. Blood vessels penetrating the yolk sac transport food to the developing animals.

(2) Internal: Mammals employ internal fertilization and, typically, their embryos develop internally within a structure called the uterus. The egg has relatively little yolk and is very small. The food source will consist of the nutrients in the blood of the mother.

(a) Placental animals: Some adaptations for animals which develop internally are highly complex and provide the protection, nutrients, and gas and liquid exchange that are essential to embryonic development. These adaptations include:

—the amnion which encompasses the amniotic fluid. This fluid provides a water environment and protection against shock.

—the placenta which is composed of tissues from both the embryo and the parent. It is the site of the transfer of oxygen, food, wastes, and other materials between the parent and the embryo. There is no direct connection between their bloodstreams; transport is accomplished by diffusion and active transport.

—the umbilical cord which is the structural and functional attachment of the embryo to the placenta.

(b) Non-placental Animals: Pouched mammals, some fishes and some reptiles employ internal fertilization and internal embryo development without direct nourishment from the parent. The source of food is the yolk stored in the egg. Adaptations for the development of these embryos are the same as in birds, except that the shell is not calcareous.

2. Post-Embryonic Development: The growth and change of the embryo to adult may be referred to as maturation. Postembryonic development also involves cell multiplication and differentiation, but the most obvious changes involve increases in size. During this process most animals grow in various directions at once at different rates, giving each one its distinctive shape. The maturing process may be essentially uninterrupted, as in the case of mammals, or it may be interrupted, as in the case of Arthropods undergoing molting or metamorphosis. Differentiation is completed when all organs have reached adult form. Further cell division is then restricted to repair and replacement.

B. Plants

1. Seed Formation: The plant embryo consists of the epicotyl, cotyledon(s), and the hypocotyl. The epicotyl develops into the leaves and

upper portions of the stem; the cotyledon(s) store(s) food; the hypocotyl develops into the root and lower portion of the stem. The embryo is enclosed in seed coats which develop from the outer coverings of the ovule. The embryo and its coats are called a seed.

2. Seed development: In flowering plants, seeds may develop inside a fruit. The fruit is one of the means of seed dispersal in plants.

Under suitable environmental conditions, seeds will germinate, producing new plants. These conditions include:

—sufficient moisture
—proper temperature
—sufficient oxygen.

The transformation of a seed into a mature plant, capable of reproduction, involves cell division, differentiation, and growth. Growth in higher plants is restricted largely to specific regions known as meristems.

Apical meristems are found in the tips of roots and stems. A number of plants also contain an active lateral meristem region, commonly known as the cambium. The cambium is not active in many herbaceous dicots. The cambium (actually the vascular cambium) should not be confused with another kind of lateral meristem—the cork cambium—present in most woody plants. These woody growth regions contain undifferentiated cells which are undergoing active cell reproduction. As a result of differentiation, various tissues are developed according to the pattern which is distinctive for each species of plant.

Apical meristems are concerned with growth in length. The cambium, located between the xylem and phloem, is responsible for the growth in diameter of roots and stems. Plant growth changes include also the formation of leaves and floral parts.

IV. REPRODUCTION AND DEVELOPMENT IN MAN

A. Reproduction

1. Gamete Formation

a. Male Reproductive System

(1) Structure: Testes are paired structures which produce the male gametes, sperm; they are located in an outpocketing of the body wall known as the scrotum. Scrotal temperature is 2–4 degrees lower than body temperature. This lower temperature is optimal for sperm production and storage. A series of ducts furnish a passageway for sperm from the testes to the outside of the body. Glands secrete a liquid into these ducts which serves as transport medium for the sperm—an adaptation for life on land.

(2) Function: In addition to producing sperm, the testes also produce male sex hormones. These regulate the development of secondary sex characteristics, such as beard development and voice quality.

b. Female Reproductive System

(1) Structure: Ovaries are paired structures located in the

lower portion of the dorsal abdominal cavity. Ovaries produce eggs by meiosis in tiny cavities called follicles. Egg cells leave the follicle of the ovary (ovulation) and are transported through the oviducts to the upper end of the uterus. At the lower end the uterus is connected to a muscular tube, the vagina. At birth, all of the eggs that a female will ovulate are present in the ovaries. Usually only one egg is released in ovulation during each cycle.

(2) Function: In addition to eggs, the ovaries also produce female sex hormones. These regulate the development of secondary sex characteristics, such as the development of the mammary glands and the broadening of the pelvis, and they also play a part in the menstrual cycle.

(a) Role of Hormones: The reproductive cycle of the human female involves the interaction of the pituitary gland, the ovary, and the uterus. Hormones from the pituitary gland control the function of the ovaries during this cycle. The ovaries, in turn, produce hormones which control the changes of the uterus. In addition, the hormones from the ovary control the production of hormones by the pituitary gland (feedback mechanism).

(b) Menstrual Cycle: The menstrual cycle consists of four recognizable stages:

—follicle stage
—ovulation
—corpus luteum stage
—menstruation

The last stage refers to the periodic shedding of the uterine lining which occurs when fertilization does not take place. The duration of each cycle is approximately 28 days. These cycles begin at puberty, are temporarily suspended during pregnancy, and end at menopause. The duration of the cycle may vary considerably from female to female and may be interrupted by illness.

2. Zygote Formation: Sperm and egg usually fuse in the upper portion of the oviduct. If more than one egg is released and fertilized, multiple births may result. Identical twins may develop from separated cells derived from the same zygote. If the egg is fertilized, cleavage begins in the oviduct, and 6 to 10 days later the resulting embryo may become implanted in the uterus. If the egg is not fertilized within approximately 24 hours after ovulation, it deteriorates and leaves the body with menstrual discharge through the vagina.

B. Development

1. *Prenatal*: Embryonic development includes:

—cleavage in the oviduct. At this time the number of cells increases without a corresponding increase in mass

—gastrulation occurs after implantation in the uterus

—differentiation, or the formation of specialized tissues and organs from the embryonic layers although no organ in the body is derived exclusively from any one of the three primary germ layers

• from the ectoderm—nervous system and epidermis of skin

• from the mesoderm—muscles, circulatory system, skeleton, excretory system, gonads, inner layer of skin.

• from the endoderm—lining of digestive and respiratory tracts, portions of liver and pancreas

—growth, an increase in cell number as well as in size of cells. Early growth consists chiefly of differentiation of cells and tissues. This is followed by changes in size and structure.

Birth usually occurs after a gestation period of nine months. Gestation is the period of time between fertilization and birth in mammals.

2. *Postnatal*: Development continues with various parts of the body continuing to grow at differing rates until the adult state is reached. Human reproductive tissues undergo prolonged development. Puberty, the onset of reproductive capability, is regulated by the nervous and endocrine systems and generally occurs between the ages of 9 and 18.

In time there is a deterioration of structure or function in the mature individual. This aspect of development is known as "aging," a result of the interplay of hereditary and environmental factors, which ultimately ends in death. The cause or causes of the degenerative changes which are characteristic of the aging process are still not fully understood.

MITOSIS

AIM: To study the process of mitosis.

Mitosis is a cell division in which the chromosomes split, doubling the number and leaving each new cell with the same number of chromosomes as the parent cell. Chromosomes are rodlike bodies, consisting of chromatin granules, in the nucleus of cells; chromosomes are known to carry hereditary traits.

Mitosis is essentially the same in principle in both plant and animal cells since there are essentially four phases in the process of mitosis.

Prophase is the first stage in mitosis when chromatin granules form spireme and then break up into chromosomes; the chromosomes then divide lengthwise in half.

Metaphase is the second phase of mitosis in which the chromosomes line up at the center or equator of the cell; they divide longitudinally and the daughter chromosomes move to the poles of the mitotic spindle.

Anaphase is the third phase of mitosis; one chromosome from each pair is moved to opposite sides of the cell.

Telephase is the terminal stage of mitosis during which the nuclei revert to a retiring stage; the chromosomes reach the poles of the cell and the cell plate begins to form.

Interphase is considered by some as a phase of mitosis but actually it is the resting stage between telephase and prophase.

Chromosomes from living cells can be studied if they are properly prepared and stained. Living onion roots can be prepared by placing an onion on top of a beaker with the bottom part of the onion submerged in water. Scrape the onion stem base before placing in the water. Usable roots should appear in about 3 days, and new ones will appear as the first ones are removed.

MATERIALS NEEDED: Onion root tip; compound microscope; microscope slide; cover glass; razor blade; forceps; Bunsen burner; pencil with an eraser; orcein; acetic acid; 1N HCl; medicine dropper; watch glass, prepared slides of onion root tip if students are having difficulty with the process.

PROCEDURE:

1. Place 9 drops of 2% aceto-orcein (2% orcein in 45% acetic acid) in a watch glass and add one drop of 1N HCl.
2. With a sharp razor blade remove about 0.5 mm from the end of a growing onion root tip and place it in the stain.
3. Use forceps to hold the watch glass while heating it gently for 5–10 seconds. See that the stain does not boil. Add 2 drops of 1N HCl and repeat the heating several times.
4. Place a drop of 1% aceto-orcein (1% orcein and 45% acetic acid) on a slide and transfer the root tip to it.
5. After placing the cover glass over the root tip, press gently with the eraser end of a pencil to spread the root tips in a thin layer over the slide.
6. With the aid of the microscope first scan the entire area under low power and then high until you find chromosomes in the dividing cells. If you have difficulty in locating, check with your laboratory partners. Someone will have a good slide.
7. Draw the various stages of mitosis as you find them.

OBSERVATIONS AND CONCLUSIONS:

1. What is the difference between mitosis in animal cells and mitosis in plant cells?
2. In what stage do the centromeres appear?
3. In what stage are there nucleoli?
4. Is the nuclear membrane present throughout the process of mitosis?
5. Summarize the different changes that are undergone by the chromosomes throughout the entire process of mitosis.
6. What is the difference between mitosis and meiosis?

REPRODUCTION IN PARAMECIUM

AIM: To study asexual and sexual reproduction in a single organism.

In your previous work with protozoa, you probably have seen Paramecium and Ameba dividing by means of fission into two new organisms. However, Paramecium also reproduces sexually as well as asexually. The sexual reproductive process is known as conjugation.

In this exercise you will study both types of reproduction as they occur in Paramecium.

If you have time, though, you might look at prepared slides of Spirogyra, a plant, and see conjugation there which is really a fusion of similar or like gametes to form a zygote which can withstand adverse conditions and when conditions grow favorable, grow back into a plant again.

MATERIALS NEEDED: Paramecium culture; methyl cellulose; slide and cover glass; iodine stain; medicine dropper; prepared slides of Spirogyra; prepared slides of Paramecium; microscope.

PROCEDURE:

1. Place a drop of Paramecium culture on a slide and slow it down with methyl cellulose.
2. Examine under the microscope and if you find a Paramecium that is dividing in two, place a drop of iodine stain on the slide so that the animal can be killed and stained for further study.
3. Make drawings of what you see in the two cells as they are being formed.
4. If you can find two paramecia which seem to be stuck together side by side, you have probably found the sexual process they undergo known as conjugation. Draw what you see.
5. Stain the slide with iodine. Carefully observe the nucleus. Is it similar to the condition of the nucleus as you saw it during the process of fission? Observe both paramecia very carefully for any other differences.
6. If unsuccessful with live Paramecium, use the prepared slides of conjugating paramecia.

OBSERVATION AND CONCLUSIONS:

1. In 1937 Tracy M. Sonneborn found that individuals engaged in conjugation apparently were different mating types; that is, actually they would correspond to what we call males and females and that organisms of Type I only conjugated with organisms of Type II. In your observations, did you find any differences in the make-up of the 2 paramecia you observed in the process of conjugation? What were they?
2. Would asexual or sexual reproduction be more conducive under adverse conditions? Why?
3. If you did look at a prepared slide of Spirogyra, would you arrive at similar conclusions in regard to sexual and asexual reproduction in plants?

REGENERATION AND REASSOCIATION

AIM: To determine how small parts of an injured organism can reform and become whole again .

Regeneration is the replacement of a lost body part, the growth of a new individual from a small piece of the organism, or the replacement of a missing limb or other major part of the body. Some examples are the hydra, starfish, planaria, and the earthworm.

It is also possible for clusters of cells to form a complete organism and this process is known as reassociation.

You will notice that the wound heals and a perfect replica of the missing part will appear from a small bud of undifferentiated cells appearing at the cut surface.

MATERIALS NEEDED: Hydra; planaria; small petri dishes; razor blade or fine scalpel; microscope slides; camel's hair brush; dissecting microscope; pond water; silicone gum; cover glasses; medicine droppers.

PROCEDURE:

(A) Regeneration in Planaria

1. Prepare 3 labeled dishes of pond water. Mark one *anterior*, the second mark *middle*, and the third mark *posterior*.

2. Transfer a planaria by means of a medicine dropper to a glass slide and with the aid of the microscope and a sharp cutting edge of the razor blade or scalpel, cut the planaria into three pieces and place the proper pieces in the proper dishes. Use several planaria cut in the same manner and place in the proper dishes.

3. Now take several planaria and cut them vertically so that they have been cut into two identical halves. Place in labeled dishes of pond water.

4. You can also cut planaria in any other manner you wish so that 2 heads are on the anterior end attached to a single body or on posterior ends. There is no end to the type of cuts that can be made.

5. Keep all the dishes in a dim light in a cool place. Observe them daily and remove any dead pieces. Change all water as soon as a dead piece is found and if there are no dead pieces, change the water every other day.

6. Keep a record of what happens by making sketches of what you observe.

(B) Regeneration in Hydra

Take several healthy hydra and cut them into several sections. Place the parts in properly labeled dishes and also observe for several weeks.

(C) Reassociation in Hydra

Place a hydra on a cover slip in a drop of pond water. Place a slide on top and gently mash the hydra. Lift the slide. Place the cells which have stuck to the cover into a living cell which you have prepared with silicone gum. Add a drop or two of pond water. Cover with the cover glass and place in a cool dim place. Observe daily for several weeks. Add water if necessary.

OBSERVATION AND CONCLUSIONS:

(A) Regeneration in Planaria: In those pieces that did not die when cut into three parts, a perfect planaria might have resulted from each of the cut parts. It is more likely that the anterior end piece would become whole again than the posterior piece. The posterior piece probably will have an abnormal head. When divided in half, it is likely you will get a perfect planaria.

(B) Regeneration in Hydra: Perfect hydra will result under the proper conditions.

(C) The cells may join together and large clumps then may form perfect hydras.

1. Which parts of the planaria had the greatest survival rate (anterior, middle, posterior, right, or left)?
2. Along which part of the cut edges did regeneration first appear?
3. Was there any difference in the appearance of the regenerated tissue from that of the old tissue?
4. Is there any rule regarding the location of the new head and new tail in planaria?
5. Is there any difference in the rate of regeneration of the various pieces?
6. Is there any difference in the capacity of various sections to regenerate?

VERTEBRATE REPRODUCTION

AIM: To study the reproductive systems of various vertebrate animals.

In this exercise you will study the reproductive systems of a fish, amphibian, bird, and mammal.

The fish and amphibian lay their eggs in the water and they are fertilized there with the yolk developing in the water and being used by the embryo as food.

The bird has internal fertilization but lays an egg with a shell surrounding the yolk. The bird embryo develops in the shell, uses the yolk as food and pecks its way out of the shell when fully developed.

The mammal has both internal fertilization and internal embryonic development.

MATERIALS NEEDED: Male and female of each of the following: fish, frog, chicken and rat; dissecting pans, dissecting materials; paper towels.

PROCEDURE: Different teams of students for each dissection and then pool the results.

(A) Fish: Make a ventral incision from the tail region to the head. Pin down the flap of the body wall so that the abdominal cavity is fully exposed.

In the male you should see two long, thin, white structures running along practically the entire length of the abdominal cavity. These are the testes and a sperm duct leads from each of the testes to a rear opening.

In the female there will be a large mass of eggs, called roe, covered by a thin transparent membrane. This mass of eggs will fill a large part of the abdominal cavity. The oviduct can be traced to a rear opening (the cloaca).

(B) Frog: Dissect your frog as you did to study the general organs. Pin back the skin flaps to expose the internal organs.

In the female there will be large masses of black and white eggs. If you lift the ovaries on one side you can see the oviduct and trace it out to the opening.

If there are no masses of eggs, lift and cut out the intestines and look for round white organs next to the kidneys. This is a male.

(C) Chicken: Open the abdominal cavity of the birds.

In the female you will find only a left ovary. By the time the bird has reached the reproducing stage, the right one atrophies and disappears or is very tiny. The ovary will look like a bunch of grapes. If you carefully examine it you will find a number of egg cells or yolks in various stages of development including some that are ready to be laid since they are completely enclosed in shell.

In the male, there are two testes which are located just to the back of the kidneys. These are white structures which are fairly long and narrow. You can trace the sperm ducts to the rear opening.

(D) Rat: Make a cut so that the abdominal cavity is exposed. Snip the large intestine and cut the esophagus where it enters the stomach. Remove the entire intestinal tract so that the reproductive system can be seen.

In the female the ovaries are whitish structures just back of the kidneys. The oviducts, uterus and vagina will appear as a Y-shaped tube leading from the ovaries. There might be embryos in the uterus. You can recognize these by the bulges. You also might see small discolored areas in the uterus. These are scars which indicate where the last embryos were attached to the uterine wall.

OBSERVATIONS AND CONCLUSIONS:

(A) Fish

1. How many testes does the male fish have?
2. How does fluid with sperm leave the body of the male fish?
3. How many ovaries does the female fish have?
4. How do eggs leave the body of the female fish?

5. How are the eggs fertilized by the sperm?

(B) Frog

1. How many testes does the male frog have?
2. How does the fluid with sperm leave the body of the male?
3. How many ovaries does the female frog have?
4. How do eggs leave the body of the female frog?
5. How are the eggs fertilized by the sperm?
6. In what way is frog reproduction similar to fish reproduction?

(C) Chicken

1. What would happen if both ovaries atrophied?
2. How is the reproductive system of the bird similar to the other vertebrate systems you have already seen?
3. How is the reproductive system of the bird different from the other vertebrate systems you have already seen?
4. Why are there yolks in various stages of development?
5. Where does the fertilization of the egg take place?

(D) Rat

1. How is the reproductive system of the rat similar to the other vertebrate systems you have already seen?
2. How is the reproductive system of the bird different from the other vertebrate systems you have already seen?
3. How is a mammal's reproductive system adapted for the survival of its young?
4. What type of epithelial tissue is found lining the oviduct of the rat? Why?

VERTEBRATE DEVELOPMENT*

AIM: To learn how different vertebrate embryos develop into fully developed organisms.

In its early stages, the embryos of all vertebrates are basically the same. The egg cell is fertilized by the sperm cell and begins an orderly process of repeated divisions known as cleavage.

These divisions continue until a hollow ball whose wall consists of a single cell layer is formed. This is called the blastula. As the divisions continue, the blastula folds in and forms a two-layered, cup-shaped mass which is called the gastrula.

Soon the gastrula forms three basic tissues, the ectoderm (outer), the

* Starred experiments throughout the book should be checked for materials at least two days in advance. Some materials must be prepared ahead of time.

mesoderm (middle), and the endoderm (inner) layer of cells. From these three tissues all the organs and systems are developed.

This development of an animal from fertilization until hatching or birth is called embryology.

In this exercise, you will study the development of a fertilized frog's egg until it becomes an adult frog as well as the development of a fertilized chick's egg, already laid, until it hatches and pecks its way out of the shell.

MATERIAL NEEDED: Developing frog eggs; watch glass; dissecting microscope; forceps; aquarium; commercial baby-food spinach; fertilized hen's eggs; incubator; culture dishes; scissors; hand lens.

PROCEDURE:

(A) Frog

In early spring, frogs lay their eggs in the water in the early morning. You will be able to recognize the fertilized eggs by looking for masses of eggs in a jelly. Bring them into the laboratory and keep them in clean pond or aquarium water. You will be able to observe the entire development. The early stages are very rapid but it takes about 6 days before tadpoles actually are hatched.

If you are unable to obtain freshly fertilized eggs from a pond, inject some pituitary tissue into the female and she will lay her eggs onto a previously induced sperm suspension in a petri or culture dish. The sperm can be removed from a pithed frog's testes and placed in about 10 ml of pond water. The sperm usually remains active enough for about 6 hours to penetrate the eggs.

1. If you are using an artificially fertilized suspension of eggs you should be able to observe something within 10 minutes. Half the egg will appear black while the rest will be white. A jellylike mass will surround each egg and there will be a random position of black and white portions.

2. About an hour after fertilization, each egg will separate from the others as the jelly swells. The eggs will rotate so that the yolk-area is facing downward.

3. About 3 hours after fertilization you will see the first step in cleavage as the single cell becomes 2.

4. Look at the eggs whenever you have a chance and at the end of about 16 hours it will be a blastula. At the end of 36 hours it will be a gastrula. In 3 days you can definitely see a neural tube and in 4 days you can begin to see a tail bud.

5. With continued daily observation you will see the heartbeat, gill formation, the mouth, and eyes.

6. In about a week begin to feed the tadpoles. Use the commercial baby-food spinach.

7. In order to observe the metamorphosis of the tadpole into a frog, the tadpoles can be kept in an aquarium filled with pond water.

(B) Chick

Fertile eggs can be obtained from local farm supply stores or local farms. There is certain to be one around in your area even though you might be living in the city. Else check with your local hospital. They use the eggs for various research projects and they might be able to obtain them for you.

The eggs you get in your grocery store are not fertile. Why?

You will observe an unincubated egg as well as one that has been in an incubator 2 days, 4 days, 10 days, and 18 days.

1. Place the unincubated egg into a culture dish. Insert the point of the scissors into the shell slightly to the right of the longitudinal axis of the egg. Cut the top off the egg by making a completely oval incision. Gently lift the top of the shell with the aid of the forceps. You will then locate the membranes that adhere to the shell. The albumen or egg white has thickened, twisted strands at each end of the yolk which are called chalazae. The yolk is very large because it is the primary source of food for the developing embryo. The embryo absorbs the albumen in later stages of development. On the surface of the yolk you will find a tiny mass of cells, called the blastoderm. These are the living cells of the embryo since cleavage has already begun.

2. Now take an egg that has been incubating for 2 days. Remove the shell from the egg as you did previously. The mass of cells on top of the egg will be an embryo about 10 ml long. If you examine the embryo with a hand lens, you will find that there is a pulsating mass which is the beginning of a heart as well as some blood vessels over the surface of the yolk. There will be paired somites in a chain which will form the backbone and skeletal muscles.

3. Then take a 4-day egg. Cut the shell as before. Draw off any albumen in the shell with a medicine dropper so that yolk surface is completely bared and study the embryo. You will now notice the heart quite clearly and there will be a relationship between pulsations of the heart and the surges of the blood. You should be able to distinguish the head and neck. There will be two bulges where the eyes will be. There will be 3 membranes. The allantois contains wastes and partially covers the embryo. The yolk sac and the amnion will enclose the embryo.

4. With the 10-day egg, be careful not to insert the scissors too far through the shell since some of the delicate membranes are to be found quite close to the shell. Describe the condition of the embryo and the various parts of it that you can recognize.

5. When you are ready to open the 18-day egg, another procedure is needed. Crack the large end of the egg with the scissors and use the forceps to pick away the shell so that no shell membranes are broken. After part of the shell has been removed place the egg into a culture dish and pick off the remainder of the shell. Now you can carefully study the embryo. It should be recognizable as a chicken with all the body parts there. You

might also find the egg tooth which is used for breaking through the egg shell. Describe the various parts of the embryo. What has happened to the size of the yolk? Has the chick been breathing air by having its beak in the air space?

OBSERVATIONS AND CONCLUSIONS:

(A) Frog

1. What is the function of the jelly around the eggs?
2. What is the dark region of the egg called? The light region?
3. Why does the dark portion of the egg rotate so that it becomes uppermost?
4. How does the tadpole hatch out of the jelly?
5. Describe the newly hatched tadpole.
6. What changes take place in the appearance of the tadpole as it becomes an adult frog?

(B) Chick

1. Why is a large yolk necessary?
2. From what is the circulatory system derived?
3. What is the amnion filled with that surrounds the chick embryo? Is this a similar environment to that in which the frog develops?
4. Does the chick embryo look like a chicken or like any bird in general?
5. By the time the chick is ready to hatch, has the yolk been completely absorbed?

POLLEN GRAINS: THEIR STRUCTURE AND GERMINATION*

AIM: To examine pollen grains, determine variations of the pollen grains in various species and to observe the germination of pollen grains and the formation of pollen tubes.

Pollen are tiny yellow bodies produced by the anthers of seed-bearing plants that contain the male sex cells. A pollen grain is an individual pollen particle, a spore usually consisting of a single cell. The pollen grains are found in pollen sacs which are sacs in the anther.

Pollination is the transfer of pollen from the male to the female part of plants. Once the pollen is transferred, a pollen tube is formed. A pollen tube is a threadlike projection which grows downward through the stigma and style from the pollen grain and conveys the male nuclei to the egg nucleus. Fertilization then follows.

When the pollen is shed from the mature anther of a stamen, the outer wall of the pollen grain has thickened and become a protective covering called *exine*. It is this outer wall that is used as a basis for plant identification. The wall develops spines, ridges, plates, pits, and other markings which become characteristic of the pollen grains of a species just as the leaves, flowers, fruits, or seeds are.

MATERIALS NEEDED: A collection of anthers from a variety of flowers; envelopes to hold the anthers; microscope; slides; cover glasses; silicone gum; methylene blue stain; dissecting needle or small camel's hair brush; Bunsen burner; sucrose (cane sugar); graduated cylinders; small beakers (6).

PROCEDURE:

When you collect the anthers, collect from as many different genera and families as possible. Use various species of grasses and if possible various types of trees including conifers.

1. Mount each kind of pollen on a different slide in a drop of water with the aid of the dissecting needle or brush. Add a drop of the methylene blue stain. Allow the stain to stand for a few minutes. If the stain is not absorbed, heat the slide over the flame by just passing the slide over, not holding it over the flame. When the pollen has been stained, add cover glass and study the various pollen grains under the low- and high-power of the microscope. Draw a variety of the pollen grains to show their characteristics. Be certain to name each pollen grain.

2. Germinating Pollen Grains: Normally pollen grains form pollen tubes when they are mature, ready to germinate, and are on the stigma of a flower pistil. However, you will attempt to germinate pollen grains in the laboratory. This usually can be done if they are placed in a sugar solution, and through experimentation, it has been found that a 10% sucrose (cane sugar) solution is best.

Prepare a 10% sucrose solution by adding 1 gram of sugar to enough distilled water to make 10 ml of the solution. Also prepare a living cell by using silicone gum. Place a drop of the sugar solution into the cell and add several pollen grains to the solution. Cover with the cover glass. Place under the low-power of the microscope and watch for the pollen tubes to emerge. It should begin in about 20 minutes. Notice that there are nuclei in the pollen tube as it grows. Draw several stages of the growth.

3. Although we used a 10% sugar solution in part 2, it does not follow that all the types of pollen will grow best in the solution. Therefore make a number of silicone gum living cells and use various types of the pollen grains in the solution. Record in table-form as to the time that it took for the pollen grain to emerge.

4. Now make up different concentrations of sugar solutions such as 5%, 15%, 20%, 25%, and 30% and try each variety of pollen in the different concentrations. Make up a table to show the time it took for the pollen tube to emerge in the different concentrations as compared with the original 10% solution.

5. If you have the time, it might make for an interesting exercise, if glucose, maltose, lactose, or any other sugar is used in place of the sucrose.

OBSERVATIONS AND CONCLUSIONS:

1. Do all pollen grains look alike?
2. If you have several anthers from flowers that are closely related, are the pollen grains similar?
3. Is there any generalization that could be made as to the correlation of concentration of the sugar solution and the time needed for germination?

FLOWERS, FRUITS, AND SEEDS

AIM: To study the various parts of the plant that are concerned with reproduction.

The flower is the organ of the plant used in sexual reproduction. The anther is the male organ and the pistil is the female organ. Through pollination, the pollen grains have landed on the stigma and by means of the pollen tube fertilize the ovules which are located in the ovary of the flower. The fertilized ovules become seeds, while the ovary enlarges and ripens to become the fruit.

We are familiar with many of these ripened ovaries, such as apples, oranges, cucumbers, pea pods, grapes, cherries.

In this exercise we shall attempt to learn more about the structure of these flowers, fruits, and seeds, and find out why they are important for the reproduction of the plant.

MATERIALS NEEDED: Soaked lima beans; germinated bean seeds both 3 and 10 days old; dissecting microscope; compound microscope; iodine solution; whole pea pods; razor blade; biuret solution; apple.

PROCEDURE:

(A) Flower

Hold the flower upright with the narrow stalk at the bottom. Locate the sepals below the petals. Observe the colorful petals. Count them. Remove the petals to expose the stamens and the pistil. Use the dissecting microscope to observe these organs. Now use a razor blade to cut along the length of the ovary to observe the tiny ovules. Surrounding the stigma and style, the anthers and filaments are to be found. Observe them under the microscope. Remove a single anther and filament and observe under the dissecting microscope. Dust some of the pollen onto a slide and observe under low- and high-power of the compound microscope. Label the parts of the flower diagram.

(B) Fruit

1. Open the pea pod carefully and remove a single pea from its stalk. Locate a tiny opening about the size of a pinhole along its surface. Also locate the point at which it was attached to the stalk. Remove the double seed coat and observe the fleshy cotyledons and embryo plant. Complete your observations and draw a labeled diagram of the pea seed.

2. The apple is an example of a fleshy fruit called a pome. Examine the stem of the apple. At the other end of the apple are to be found the remains of the sepals, and a ring of dried stamens. Cut the apple in half lengthwise. Notice the various parts of the apple. The papery core is the ovary wall within which are the seeds attached to a central placenta which runs through the ovary. Draw a labeled diagram of the cut apple.

(C) Seed

1. Observe the lima beans. They have been soaked in order to soften their tissues and thus make them easier to examine. Find the pinhole-sized opening on the surface of the bean. Directly above this hole locate the scar that remains where the bean was removed from its point of attachment in the bean pod. Remove the double seed coat carefully. Separate the two halves of the beans to expose the embryo plant carefully. Use the dissecting microscope to examine the embryo plant carefully. Draw and label the large flat structures, the cotyledons, and the embryo plant structures, the epicotyl. Observe that these structures arise from the point at which the embryo is attached to the cotyledons. Test one cotyledon for starch with the iodine and for protein with the biuret solution.

2. Take a bean seed that has been germinating for three days. Observe that there is a root. What part of the embryo gives rise to the root? Where are the first leaves of the seedling? What part of the embryo produces the first leaves? Test the cotyledons for starch and protein.

3. Then observe a bean seed that has been germinating for 10 days. What has happened to the cotyledons? Where is the seed coat? What part or parts of the embryo developed into the stem? Also, how are the first two miniature leaves arranged on the stem?

4. Draw the seed in parts 1, 2, and 3 of this exercise.

OBSERVATIONS AND CONCLUSIONS:

1. Define the following: complete flower, calyx, corolla, stamens, pistil, incomplete flower, perfect flower, imperfect flower, staminate flower, pistillate flower, stalk, sepals, anthers, filaments.

2. How do the petals differ from the sepals?

3. What is fertilization?

4. After fertilization, how does the seed form?

5. How does the fruit form?

6. How are the garden pea flower and the pea pod related?

7. List two differences and two similarities between the pea and the lima bean seeds.

8. What is the function of each part of the pea or bean seed?
9. What parts are essential for germination?
10. How does the plant embryo use the cotyledon?
11. In the pea pod, why does it happen that the pod might contain undeveloped seeds?
12. How do the plant embryos use the cotyledon?
13. How does man use the cotyledons?
14. What are the functions of the fruit?
15. Can you find the various parts of the flower when you examine the fruit? Describe in the apple.
16. What are the functions of the fruit?
17. What becomes of the cotyledons during germination?
18. What functions do some fruits serve for man?
19. What fruits are not eaten by man?
20. What are the various agents used in seed dispersal?

ENZYMES AND GERMINATING SEEDS

AIM: To show that germinating seeds have more enzymes that split starch than do dry seeds.

We know that for a seed to germinate and to develop, it needs food. Usually this food is found in seeds as starch. But the seeds cannot live by starch alone and therefore starch must be split into sugar, such as maltose and glucose. The enzyme that acts on the starch is only activated when water has been absorbed by the seed.

Starch is present in the seed in the form of insoluble grains of starch and it is the digestion of these grains which permits the glucose and maltose to be used in respiration and in the growth of the seed to a plant. There are a number of enzymes involved in this change. The starch is broken down into maltose by means of beta-amylase while the maltose is broken down to glucose by means of maltase.

In this exercise we shall see that the enzymes are only activated when water is absorbed and under no other circumstances. We shall use iodine to show the presence of starch and Tes-Tape to show the presence of sugar.

MATERIALS NEEDED: Fifteen corn grains; 3 petri dishes containing starch agar; razor blades; iodine solution; Tes-Tape; beaker; water; tripod; iron gauze; Bunsen burner.

PROCEDURE:

This exercise can be done in teams.

1. Take ten of the corn grains and soak them for 24 hours.

2. The next day, cut 5 of the corn grains in half lengthwise and boil the 10 halves for about 30 minutes. Cool.

3. Meanwhile cut in a similar fashion the 5 dry grains, and the 5 remaining soaked corn grains.

4. Label the 3 petri dishes A, B, and C.

5. Into "A" place 5 half-grains of the dry grains so that the cut edge is in contact with the surface of the agar.

6. Into "B" place 5 half-grains of the soaked corn grains, so that the cut edge is in contact with the surface of the agar.

7. Into "C" place 5 half-grains of the corn that was boiled so that the cut edge is in contact with the surface of the agar.

8. Allow the petri dishes to remain at room temperature for 24 hours.

9. Remove the corn grains from the dishes. Pour a few drops of iodine solution into each of the petri dishes and swish around so that the entire surface of the agar has been covered by the iodine solution. Pour off the excess iodine and pour several drops of tap water onto the surface, swish and pour off. Note the results in each of the dishes. If the starch has been digested, there will not be a blue-black color where this has taken place.

10. Dip one end of the Tes-Tape paper into the area where the corn grains were placed and allow to remain for about 10 minutes.

11. Note any changes in the color of the paper.

OBSERVATIONS AND CONCLUSIONS:

1. What did you find when you flooded the petri dishes with the iodine solution? Did all three show enzyme activity?

2. Did you expect these results? Explain why.

3. When you tested the area where the embryos had been placed with the Tes-Tape paper, what did you find?

4. What would you say concerning the degree of enzyme activity in a developing seed as compared to a dry seed and a boiled seed?

5. What do you think occurred when you boiled the seed?

UNIT VII

TRANSMISSION OF TRAITS FROM GENERATION TO GENERATION

I. MENDELIAN HEREDITY

Genetics, the science of heredity, is based on the principles developed by Gregor Mendel. Using garden peas, Mendel made many crosses and charted the patterns of inheritance of plant characteristics from parents to offspring. Analysis of this data permitted Mendel to devise a hypothetical model of the pattern of heredity in the pea plant. The principles of DOMINANCE, SEGREGATION, and INDEPENDENT ASSORTMENT were part of the contributions Mendel made to the understanding of the hereditary process.

A. *Dominance*: When pea plants having alternate forms of a contrasting character were crossed, the offspring resembled only one parent, the dominant one. The character which did not appear among the offspring was called the recessive character. Offspring of this type were designated as the F_1 generation.

B. *Segregation*: When F_1 individuals were crossed (self-pollinated), the hidden recessive character segregated out and appeared in about one-quarter of the offspring. These offspring were designated as the F_2 generation.

C. *Independent Assortment*: In crosses where two traits were studied at a time, they were found to be inherited independently of each other.

II. THE GENE-CHROMOSOME THEORY

After Mendel's work (which was not immediately recognized and accepted) was rediscovered, considerable investigation followed which led to the statement of the gene-chromosome theory of heredity. Among the techniques employed were microscopic examination of dividing cells and breeding experiments with the fruit fly, *Drosophila.* The hereditary mechanism proposed as a result of these investigations included the following understandings.

- The inheritance of traits closely correlated with the behavior of chromosomes in sexually-reproducing organisms. The chromosomes contain the genetic determiners in linear sequence.
- Each determiner (gene) has a definite position in the sequence of genes on a chromosome.
- Homologous pairs of chromosomes thus have paired genes in the same positions in the gene sequence.
- These pairs are known as alleles or allelic genes.

The proposal of the gene-chromosome theory made it possible to redefine some of the observations made by Mendel.

A. *Dominance*: If the two genes of an allelic pair are unlike, the individual possessing this gene combination is said to be heterozygous for the trait with which the genes are associated. If the two genes of the allelic pair are identical (either dominant or recessive), the individual possessing this combination is said to be homozygous for that trait. When an organism is heterozygous for a specific trait, the effects of the dominant gene are observed and those of the recessive gene are not. If one organism is homozygous dominant and another is heterozygous, both would exhibit the dominant trait. The two organisms would have the same phenotype but different genotypes.

B. *Segregation and Recombination*: During gamete formation there is a random segregation of chromosomes due to the events of meiosis. As a result of gametic fusion, new chromosome combinations (consequently, new allelic gene combinations) are likely to be produced.

C. *Independent Assortment and Recombination*: If the genes for two different traits (nonallelic genes) are located on the same chromosome, they are said to be linked and are usually inherited together. Due to the events of meiosis and random gametic fusion, nonlinked genes are inherited independently.

III. PATTERNS OF HEREDITY

A. *Incomplete Dominance*: Incomplete dominance involves an apparent "blending" of the effects of contrasting alleles. (Consequently, heterozygous individuals do not resemble individuals with either of the homozygous combinations. Four-o'clocks, Andalusian fowl, and roan cattle exhibit incomplete dominance for some traits.

B. *Sex Determination*: Diploid cells of many organisms contain two types of chromosomes, autosomes and sex chromosomes. In man there are 22 pairs of autosomes and one pair of sex chromosomes. The sex chromosomes are designated "X" and "Y". The sex of an individual is determined at fertilization on the basis of the resultant combinations of X and Y chromosomes.

- In humans and fruit flies, XX is the condition in females and XY the condition in males.
- In other organisms, such as domestic fowl, males possess two identical sex chromosomes (ZZ) and females two unlike sex chromosomes (ZW).

C. *Sex Linkage*: Genes for certain traits are located on the X chromosome. These traits are said to be sex-linked and do not appear to have corresponding alleles on the Y chromosome. Hemophilia and color-blindness are examples of sex-linked traits in humans.

D. *Multiple Alleles*: In some instances the observed pattern of heredity cannot be explained satisfactorily on the basis of a single pair of contrasting alleles. It has been possible to devise a satisfactory model in many instances by assuming that more than two alleles are involved in the inheritance of particular traits. Such a model is said to involve multiple alleles.

In humans, the inheritance of the ABO blood groups is explained with a model that employs three alleles (I^a, I^b, and i) as follows:

Blood Group	*Genotypes*
A	I^aI^a or I^ai
B	I^bI^b or I^bi
AB	I^aI^b
O	ii

IV. MUTATIONS

Mutations are genetic changes which are heritable. Mutations that do not affect the sex cells cannot be passed on to offspring by sexual reproduction. However, mutations that occur in autosomes may be perpetuated within an individual as a result of mitotic cell divisions. Most mutations are recessive and are disadvantageous to the organism.

A. *Types*:

1. Chromosomal Mutations: A chromosomal mutation is a change in a chromosome structure or in the number of chromosomes. The effects of chromosomal mutations are often visible in the phenotype of an organism because many genes are usually involved.

a. Crossing-over: At times, during synapsis in the first meiotic division, chromatids of homologous chromosomes overlap at one or more points. When disjunction occurs, portions of one chromatid may be exchanged for corresponding portions of another chromatid. Since different alleles may be transferred, chromosome composition is changed.

b. Nondisjunction: Sometimes entire chromosomes fail to separate from one another during disjunction. This results in gametes with more (or fewer) than the normal *n* chromosome number. After gametic fusion, new individuals may have more, or less, than the normal *2n* chromosome number. Example: in humans, Down's syndrome (Mongolian idiocy) results from nondisjunction of chromosome number 21.

c. Polyploidy: Occasionally meiosis fails to occur during gamete formation. The resulting *2n* gamete sometimes succeeds in fusing with a normal *n* gamete, producing *3n* individuals. (If two *2n* gametes fuse, a *4n* individual results.) Such individuals often exhibit exaggerated characteristics such as those in "double" zinnias.

d. Other Chromosomal Mutations: Other unusual alterations in chromosome composition may result from random breakage and recombination of chromosome parts during the early stages of the first meiotic division. Recently, evidence has begun to accumulate that suggests that the hallucinogenic compound LSD promotes chromosomal breakage, hence, heritable mutations.

2. Gene Mutations: A gene mutation involves a change in the chemical nature of a gene. Often the effects of gene mutations are subtle and not readily observable. In some instances, however, gene mutations may be dramatic, as in the case of the gene which determines albinism.

B. Mutagenic Agents: Mutations may be caused by such agents as:

- Cosmic rays
- X-rays
- Ultraviolet rays
- Radiation from radioactive substances
- Compounds such as mustard gas or colchicine

V. INTERACTIONS OF HEREDITY AND ENVIRONMENT

The environment plays an interacting role in the development and expression of inherited traits. The relationship between gene action and environmental influence has been studied in many organisms.

VI. HEREDITY AND MAN

A. *Human Genetics*: The principles of genetics appear to hold for all organisms including man. Acquisition of knowledge of human genetics has been limited because man is a difficult subject for genetic experimentation. A prolonged period of physical development and a sparsity of offspring are two of the problems encountered. Some human traits that have been studied include phenylketonuria, sickle-cell anemia, and blood group inheritance.

1. Phenylketonuria: Phenylketonuria (PKU) is a condition characterized by feeblemindedness and has been shown to occur when an individual is homozygous with respect to a particular mutant gene. The symptoms of the disease apparently result from the gene-induced inability to synthesize a single enzyme necessary in the metabolic activity associated with proper brain function.

2. Sickle-cell Anemia: A condition known as sickle-cell anemia is common among certain African populations. Afflicted individuals inherit an abnormal hemoglobin which imparts the peculiar "sickle" shape to red blood cells. Sufferers of the disease often die at an early age.

Heterozygous individuals sometimes show mild symptoms of the disease. In addition, heterozygous individuals have a higher than normal resistance to malaria. This relationship has tended to maintain the otherwise harmful gene in African populations.

It has been determined that the abnormal hemoglobin differs from normal hemoglobin by only a single amino acid.

3. Blood Groups: The inheritance of the ABO blood groups in man is best explained by assuming that three alleles are involved in the expression of the trait, no individual possessing more than two of the three.

B. *Plant and Animal Breeding*: Principles of genetics have been used by man to improve and produce new varieties of plants and animals. Artificial selection, inbreeding, hybridization, and the maintenance of desirable mutations by vegetative means (grafting, for instance) are widely used.

LAW OF CHANCE

AIM:

1. To illustrate how the Law of Chance operates.
2. To show how the Law of Chance may explain the results obtained in the F_2 generation of a breeding experiment.

MATERIALS: Forty red and forty white beans of the same type mixed in each of two containers. Navy beans and red kidney beans may be used.

PROCEDURE:

Four students work together as a team. The first and second students will select and tally the first forty pairs of beans as directed below. The third and fourth students will continue selecting and tallying the second forty pairs of beans.

Using both hands at the same time (without looking into the containers), select one bean from each container. If two red beans, RR, are picked at one time, tally them by a diagonal stroke in the proper column below, and *return the beans to their respective containers.* If a red and a white bean are selected, RW, tally them and return the beans to their containers. Similarly, if you select two white beans, WW, tally them and return the beans to the containers from which they were picked.

Continue to remove the pairs of beans until 40 pairs have been tallied.

Total the number of each combination selected. Compute the percent of the 40 selections that each total represents.

$$\text{For example: } \frac{\text{Total RR}}{40} \times 100 = \text{percent RR}$$

Each team will record the total of the first forty selections; the third and fourth students will continue selecting bean pairs. Forty additional selections should be made and tallied. Indicate the totals of the second set of forty pairs. Then compute the percentage obtained for each combination out of eighty selected pairs.

$$\text{For example: } \frac{\text{Total RR (first 40)} + \text{Total RR (second 40)}}{80} \times 100 = \text{percent RR out of 80}$$

OBSERVATIONS AND CONCLUSIONS:

	RR	RW	WW
Tally (1st 40)			
Totals (out of 40)			
Percent (out of 40)			
Total (second 40)			
Total (of 80)			
Percent (of 80)			
Class totals			
Class percentages			

1. Compare the percentages of each combination obtained out of the first forty selections with those obtained out of the eighty pairs selected. Are they similar? Which percentages are closer to those obtained from the class totals? (about 9×40 or 360 pairs)? Why?

2. Under what conditions do the selections distribute purely by chance? (Consider number of selections; size of beans of each type; texture of surface of each color bean; objectivity of experimenter selecting beans.)

3. What would the percentages be if the Law of Chance completely and perfectly operated in this experiment?

4. What possible explanations for the failure to obtain these "ideal" or theoretical percentages can you offer?

5. Recalling what you have learned thus far in the study of genetics and heredity answer the following questions:

a. What is the significance of the Law of Chance in the offspring of hybrids?

b. What is the significance of equal numbers of red and white beans in each of two containers?

c. What is the significance of the *pairs* of beans of various combinations?

d. What is signified by the selection (by chance) of one bean from each container, therefore producing a particular combination?

6. When chance selections are made from each of two containers holding equal numbers of two kinds of beans (or genes) the possible combinations sort out according to the Law of ......

7. The ...... the number of selections made, the more nearly the actual results approach the expected percentages predicted by the ...... of ......

CORN: PURPLE vs. NON-PURPLE

AIM: To study purple and non-purple corn in relation to Mendelian laws. Since purple and non-purple corn are usually associated with smooth and wrinkled kernels as well, we shall attempt to count both at the same time and determine which of the Mendelian laws are in effect in this exercise.

MATERIALS NEEDED: Corns with purple, non-purple, smooth and wrinkled kernels, straight pins.

PROCEDURE:

1. Take an ear of corn and count the number of purple and non-purple kernels in each row. In order to prevent the same row from being counted twice, place a pin at the end of each row as you finish counting the row. Record your findings on a chart, listing the number of purple in each row and the number of non-purple in each row. How many of each have you found on your ear of corn?
2. Do the same thing for the same ear of corn but now count the number of smooth as against the number of wrinkled kernels of corn. Does the amount differ greatly from what you counted in step #1?
3. Now with this ear of corn, count the number of purple-smooth, purple-wrinkled, non-purple-smooth, and non-purple-wrinkled kernels. Record your findings.
4. Take a second ear of corn and do the first three steps over again.

OBSERVATIONS AND CONCLUSIONS:

1. You now have 6 tables of information. Can you make any sort of conclusion from your findings?
2. Which of Mendel's principles were used in this exercise?
3. On the basis of your findings, what was the genotype of the parents of both ears of corn? Use P for gene purple and p for gene non-purple, S for gene smooth and s for gene wrinkled.
4. Do you think that the genes for kernel color and the genes for coat texture were on the same chromosome? If they were would this cause any change in your ratios?

VARIATION AMONG LIVING THINGS

AIM: How do traits vary among living things?

We all know that the traits of an individual are the result of genetic as well as environmental factors. However, it often is difficult to determine which of these two factors is more important for the appearance of a particular trait. In this exercise, you will study two different traits, one of which is controlled by genetics and the other by environment as well as genetics. It will be your problem to determine which is which.

MATERIALS NEEDED: Twenty-five black watermelon seeds; a metric plastic ruler; a tape measure; graph paper.

PROCEDURE:

1. Measure each of the seeds with the plastic ruler and record your findings for each. Indicate the size to the nearest millimeter marking.
2. Use the tape measure to measure the wrist diameter of the members of your class. Place the information on a sheet of paper, listing separately the diameters of boys and girls.
3. Graph the results of the 3 measurements.

OBSERVATIONS AND CONCLUSIONS:

1. Which of the variation measures is primarily genetic in origin?
2. When you studied the graphs, were they similar? Why?
3. What reasons can be given for the variations you did find in the size of the wrists and in the size of the seeds?
4. Are there any other animal and plant traits that would be similar?

HUMAN HEREDITY

AIM: To determine the frequency of certain human traits in a population.

Although human heredity is studied primarily by means of pedigrees or family records, there are several traits that can be studied immediately to help you understand human heredity. The four are tongue rolling, PTC tasting, ear lobes, and eye color.

MATERIALS NEEDED: PTC (phenylthiocarbamide) papers; your immediate family, friends, and classmates.

PROCEDURE:

(A) Inheritance of Eye Color:

Make a survey of 50 persons to determine their eye color. Dark eye color consists of brown or black pigments while light consists of gray, blue, or green. Record your findings as to color and sex.

Would you say that sex of an individual is related to eye color?

(B) Inheritance of Ear Lobes:

Make a survey of 50 persons to determine whether their ear lobes are free or attached to the jaw. Record your findings as to type of lobe and sex.

Would you say that the sex of an individual is related to ear lobe attachments?

(C) Inheritance of Tongue rolling:

Make a survey of 50 persons to determine their ability to roll the tongue or their inability. Record your findings as to ability to roll tongue and sex.

Would you say that the sex of an individual is related to the ability to roll the tongue?

(D) Inheritance of the Ability to taste Phenylthiocarbamide:

Make a survey of 50 persons to determine the ability of an individual to taste PTC as a bitter sensation as against the inability to taste the PTC. It is said that about 70% of the population can taste the PTC. Record your findings as to the ability to taste PTC and sex.

OBSERVATIONS AND CONCLUSIONS:

1. Judging from the results that you have tabulated in the four various traits, what would you say were the dominant and recessive traits in each category?

2. Since you have used the members of your family for the various tabulations, would you say that two parents who are nontasters could have a child who is a taster? Would it be possible for two different children of the same parents to have different traits from each other?

3. Have you come close to the 3:1 ratio that Mendel found in making monohybrid crosses with simple dominance? If you did not, is there any explanation?

4. Are you able to work out your own genotype for the four traits we studied? Also for the immediate members of your family?

PLANT HEREDITY*

AIM: To study the relative effects of heredity and environment on plants.

Variety is characteristic of plant and animal populations. Variation is the differences of structure or function between individuals of the same species.

* Starred experiments throughout the book should be checked for materials at least two days in advance. Some materials must be prepared ahead of time.

Usually within the species a subdivision, called variety, is based on some hereditary difference which is considered too small to serve as a basis for a new species.

Many of these variations are caused by mutations which are changes in the structure or composition of a gene which is heritable. One of the mutations found in plants results in albinism which is the lack of development of normal pigments in an organism.

Although albinism in animals may be an advantage, it certainly is a detriment to plants. The lack of chlorophyll in a green plant will result in death of the plant since it cannot carry on photosynthesis.

MATERIALS NEEDED: One hundred corn seeds mixed for green and albino plants; soil flat for 100 seeds; 100 mixed tobacco seeds; 2 petri dishes; blotting paper; scissors; forceps; medicine dropper; hand lens; box to cover petri dishes.

PROCEDURE:

1. Plant the 100 seeds of corn in the flat, placing 10 seeds in a row and making ten rows. Allow them to grow as you would any box of seeds. It should take about 10 days.
2. Meanwhile cut out 2 disks to fit snugly into the bottom half of a petri dish. Divide each disk into quarters by drawing lines across. Mark one disk A and one disk B.
3. Place one disk into the bottom half of each of the 2 petri dishes. Pour water into each dish. When the paper is thoroughly soaked, pour off the excess water.
4. Sprinkle 24 tobacco seeds into each dish. Use the forceps to arrange it so that there are 6 seeds in each quarter of the dish. Allow enough space between the seeds.
5. Place the covers on the dishes. Put both dishes in a warm place that receives strong light but cover dish B with a box to keep it in darkness.
6. Check the dishes daily to make certain that there is enough water in the dishes; if not, add water by means of a medicine dropper.
7. When about half the seeds have germinated, examine them with a hand lens. Each plant will consist of a white or colorless root as well as two tiny leaves, the cotyledons. Some will have green cotyledons while some will have creamy, white ones. Count the number of each of the seedlings and record under the heading the first day. Examine both dishes.
8. Do the same thing the second day. Record any seedlings that may have died even if they had been growing the previous day. At this time allow dish B to remain in the sun.
9. On the third day count in the same manner as before. Still keep dish B in the sun.
10. Count the fourth day. Calculate the percentages of albinos in each dish.

11. At the end of 10 days, count the number of plants that have remained growing even though most of them may have germinated previously.

OBSERVATIONS AND CONCLUSIONS:

1. In the corn seedlings, what was the ratio of white to green seedlings?
2. What is happening to the white seedlings? Why?
3. What is a lethal gene?
4. When you compare the percentage of albinos in dish B on the fourth day with that on the second, what has taken place? Why?
5. How can you account for the difference among seedlings in dish A?
6. Would you say that environment or heredity plays a bigger role in albinism in tobacco seedlings?

INHERITANCE IN DROSOPHILA

AIM: (1) To learn how to culture and cross fruit flies; (2) to study their inherited characteristics; (3) to demonstrate Mendelian laws by mating special types of flies.

Drosophila are the best subjects for study of problems of inheritance in a classroom since they have a complete life cycle of about 2 weeks. Their dominant and recessive traits are very distinct and they show many mutations.

MATERIALS NEEDED: Large-mouth jars about half-pint in capacity; cotton for plugging the mouth or strips of cheesecloth to place over the open top; banana; knife; cultures of wild type drosophila, culture of vestigial-wing type drosophila; oven or double boiler for sterilizing culture jar; baker's yeast culture; ether; camel's hair brush; paper toweling; fermenting fruit; nutrient agar; cornmeal; molasses; distilled water; petri dishes.

PROCEDURE:

(A) To Collect and Grow Drosophila or Fruit Flies:

1. Prepare the bottles for the flies by boiling them for 20 minutes. Also sterilize the cheesecloth or cotton.
2. Wash your hands as well as the outside of a ripe banana.
3. Sterilize a knife blade with hot water or a flame. Peel the banana but do not touch it with your hands. Cut it with the sterilized knife, sterilizing the blade frequently as you go along. Cut the pieces of the banana directly into the bottles and then close the top. Use either cotton or a double thickness of the cheesecloth, closing the cloth tightly with string.
4. Go to any fruit store, market, or grocery and place an open culture jar near the fruit stands or fruit baskets and leave it open. The flies will enter and can be trapped.

5. Keep no more than about 10 of the flies in one jar. If you have more than that, place an empty jar against the open mouth of the jar with the flies. Have a source of light directly shining on the empty jars. Since the flies are attracted to light, they will fly into the empty jar. As soon as you have as many as you need, make certain that the jars are resealed. Keep at 75°F.

6. The flies will lay eggs on the banana and within a few days you will see maggots on the banana which will crawl up the side of the bottle and pupate. These pupae look like brown seedlike bodies. Within another week the flies will emerge from the pupae and shortly you will have enough flies to work with.

(B) Preparation of Media:

You should prepare a culture medium containing agar if you are planning a long-range study of heredity. The one that is suggested as best is prepared as follows: Add 15 grams of nutrient agar to 570 ml of water and bring to a boil. Slowly add 100 grams of cornmeal to the boiling solution. Make certain that the mixture is constantly being stirred as the cornmeal is added. Then add 135 grams molasses and boil for 5 minutes. Place in sterile jars. Keep culture medium refrigerated until used. Just before using, add a little yeast powder to several drops of water and sprinkle over culture medium to provide larvae with food.

(C) Etherization and Sorting of Flies:

In order to transfer, count, or examine the flies, they must be etherized. Use the same type of jar as you have for the cultures but place a nail through a cork that fits the top. On the side of the nail that protrudes through the cork into the bottle itself, wrap thick layers of absorbent cotton around the nail and tie with a string. Put a few drops of ether on the cotton when ready to use. Before that, with the aid of the light source, transfer several flies from the culture bottles to the empty bottle. Thrust in the cork so that the cotton is in the jar and the flies will be etherized in about 10 seconds. As soon as the flies fall to the bottom of the jar, remove them to a clean white card for study under a dissecting microscope.

A fly is dead if the wings are standing out at an angle. Use only etherized flies and they will recover within 5 to 10 minutes so that you must work very quickly in your assorting. You can reetherize them if you have ready a lid of a petri dish lined snugly with a piece of filter paper. Just put a few drops of ether on the paper and turn the dish over on the flies so that the paper is nearest them.

As you assort the flies according to their characteristics with the aid of a camel's hair brush, have ready several fresh bottles of culture medium so that the flies can be placed in the dishes accordingly.

(D) To Show Dominance:

1. Remove about 10 flies from a jar of wild-type drosophila originally

collected and grown. Anesthetize them and examine under a dissecting microscope with a hand lens to find several males. The males have a clump or patch of dark bristles on the first legs. Place these males in a tiny cone twisted from clean paper and drop the cone and flies into a fresh culture jar.

2. Now repeat this procedure by using a culture of vestigial-wing drosophila, selecting the females this time. Use a culture so as to get the females within a few hours after they emerge from the pupae. The females must be virgin. Place the females in with the males. Do not disturb the jar for about a week. By this time the pairs will have bred and there will be tiny maggots in the jar. In another week there will be pupae. At this time remove the parent flies from the jars. The flies that emerge from the pupae will be F_1 generation which is all hybrid for the long- and vestigial-wing character. The normal, or long-wing, is dominant, therefore F_1 will show all long wings. The wild type is dominant, so that the F_1 flies have yellowish-gray bodies as well as red eyes which are found in the wild type and are dominant over the brown, black and white eyes.

(E) To Show Segregation:

From the hybrids just cultured in (D) take about 10 adult flies. These flies should be just a few days old. Transfer them to a new culture jar to produce a second generation. When the pupae appear on the jar, remove the parent flies. When you have too many flies kill some of them off by over-anesthetizing them. Be certain to shake those from the jar that appear to be sluggish, because they are vestigial-winged and do not move about as easily as do the flies with normal-type wings.

Now count the offspring. You should find that the ratio of long wings to normal will be 3:1 in the F_2 generation.

If you found flies with other colors and sizes of wings, you probably have found some mutations.

OBSERVATIONS AND CONCLUSIONS:

1. Why are the Drosophila used so extensively in genetic studies?

2. When the F_2 flies emerge, how many combinations of wing length and eye color can you find?

3. Summarize the life cycle of drosophila. If possible, give the duration of each stage. Make a labeled drawing of the life cycle.

4. What do the abdomens of the females and males look like?

5. Why do we use only virgin, unmated females for our breeding experiments?

SEX CHROMOSOMES*

AIM: To study the inheritance of sex and sex-linked traits.

As we know sex is the distinction between male and female where the male produces sperm and the female eggs. In each organism there is a definite number of chromosomes but there is one set of chromosomes that is different for the male and the female. This pair determines the sex of the individual. In the female the pair is composed of two similar chromosomes, known as X chromosomes, while in the male the pair of chromosomes consists of an X and a Y chromosome. Therefore the XX is the genotype of the female and the XY is the genotype of the male.

The X chromosome carries the genes for several traits and the genes are called sex-linked genes.

Since we cannot experiment with humans to show how sex-linked characteristics work, we shall use another method of explaining the phenomenon.

MATERIALS NEEDED: White pipe cleaners; 6-inch test tubes; 10% hydrochloric acid solution; 1% sodium hydroxide solution; very dilute bromthymol blue solution.

PROCEDURE:

1. The pipe cleaners will represent the chromosomes and the test tubes the cells of the body.

2. Soak the pipe cleaners that will represent normal vision females with normal vision chromosomes in a 10% hydrochloric acid solution for about an hour. Call them CC.

3. Soak the pipe cleaners that will represent normal vision females with genes for color blindness in a 1% sodium hydroxide for about 1 hour. Call them Cc.

4. For your male chromosomes, bend the pipe cleaners to appear like a cane. They should be left untreated.

5. Prepare a very weak solution of bromthymol blue by taking the amount of bromthymol blue adhering to a wet toothpick when the toothpick is placed into it, and dissolving it in 250 ml of hot tap water.

6. Fill 4 test tubes 2/3 full of the solution. This solution acts as an indicator to demonstrate the masking of the genes when a recessive and a dominant one are present in the same cell.

7. Label tube #1 as CC and place in it two cleaners which have been soaking in the hydrochloric acid. Label tube #2 Cc and place in it one pipe cleaner soaked in the HCl and one pipe cleaner soaked in NaOH. Label tube #3 CY and place in it one pipe cleaner soaked in HCl and one cane-shaped cleaner. Label tube #4 cY and place in it one pipe cleaner soaked in the sodium hydroxide and one cane-shaped cleaner.

8. Since the solution turns yellow in the presence of an acid and blue in the presence of a base, it can easily be observed and you can determine what would happen when the various combinations of genes occur when a trait is sex-linked.

9. Tube #1, CC, turns deep blue which means that the color vision will be normal and the individual will be female.

Tube #2, Cc, turns deep blue; the color vision will be normal and the individual will be female but a carrier of the gene for color blindness.

Tube #3, CY, will turn deep blue; the color vision will be normal and the individual will be male.

Tube #4, cY, will turn yellow; the individual will be color-blind and male.

10. If you have time work out a cross between a color-blind male and a female who is a carrier of the color-blind gene.

OBSERVATIONS AND CONCLUSIONS:

1. What percent of the males in the experiment are color-blind? have normal color vision?
2. Can females ever be color-blind? If so, under what circumstances?
3. Why is it that more males are color-blind than females?

UNIT VIII

MODERN GENETICS

In recent years, biochemists have found that the nucleic acid DNA (deoxyribonucleic acid) is the chemically active component of the gene. (RNA is the genetic material of some viruses.)

A. DNA AS THE HEREDITARY MATERIAL

1. *Evidences*: DNA has been demonstrated experimentally to be the chemical basis of heredity.

- Experimental evidence involving bacterial transformations indicates that isolated DNA can bring about hereditary changes usually associated with the gene.
- Evidence from research involving bacteriophages (bacterial viruses) has also pinpointed DNA as the hereditary material.

In order to understand the chemical nature of the gene, the chemical structure of DNA must be understood.

2. *DNA Structure*: DNA, usually found in the nucleus of cells, is a large molecule consisting of thousands of smaller units known as nucleotides.

a. Nucleotides: A DNA nucleotide is composed of three parts:

- phosphate
- deoxyribose (5-carbon sugar)
- nitrogenous base (adenine, thymine, guanine or cytosine)

Adenine and guanine are purines, and thymine and cytosine are pyrimidines.

b. Watson-Crick Model: Watson and Crick, using x-ray diffraction data supplied by Wilkins, developed a model of the DNA molecule constructed of nucleotides. This model helps to explain how DNA replicates and how genetic continuity is maintained.

• The proposed DNA molecule has a "ladder" type organization. This "ladder" is thought to be twisted in the form of a double helix.

• The uprights of the "ladder" are composed of alternating phosphate and deoxyribose molecules.

• Each rung is composed of pairs of bases held in association by relatively weak hydrogen bonds. Base pairs are restricted to adenine-thymine and guanine-cytosine combination.

B. GENE ACTION: The two principal actions of genes are:

• maintaining genetic continuity by means of replication.

• controlling cellular activity by producing specific enzymes.

1. *DNA Replication*: DNA is believed to replicate in the following manner during the processes of mitosis and meiosis:

• Double-stranded DNA unwinds and "unzips" along the weak hydrogen bonds between the base pairs.

• Free nucleotides are incorporated, in sequence, forming two double strands of DNA which are identical to each other and to the original DNA molecule.

2. *Gene control of cellular activities*: The control of cellular activities by DNA involves the nucleic acid RNA (ribonucleic acid).

a. RNA: RNA, like DNA, is composed of nucleotide building blocks. Three major differences between the structure of DNA and RNA are:

• In RNA, ribose is substituted for deoxyribose.

• In RNA, uracil is substituted for thymine

• RNA is single-stranded but DNA may be either double- or single-stranded, depending on the phase of its activity. Messenger RNA and transfer RNA are two functionally different types of RNA.

b. RNA and RNA codes: It is postulated that there is a genetic code which has a four-letter alphabet:

• DNA—A. T. C. G.

• RNA—A. U. C. G.

Both DNA and RNA codes are based on sequences of three of these letters or multiples of three letters.

c. Protein synthesis: DNA serves as a template for the synthesis of messenger RNA from free RNA nucleotides. The messenger RNA molecules, carrying a specific code determined by the base sequence of the original DNA molecule, move from the nucleus to the cytoplasm where they become associated with ribosomes.

The messenger RNA is "reverse copy" of the DNA which produced it.

d. The one-gene–one-enzyme hypothesis: A current hypothesis, called the one-gene–one-enzyme hypothesis, proposes that the synthesis of each enzyme in a cell is governed by the action of a single gene.

e. Individuality of organisms as related to their RNA and DNA codes: Since the code of DNA determines the code messenger RNA, DNA determines the sequence of the amino acids in specific proteins. The specificity of enzymes is dependent on their protein make-up, and, since the individuality of a cell is largely a function of the enzymes it possesses, it is evident that DNA determines the individuality of an organism.

C. GENE MUTATIONS

1. *Characteristics*: Gene mutations may be interpreted biochemically as any change that affects the base sequence in an organism's DNA.

2. *Mechanisms*: Gene mutations include additions and deletions of bases in the DNA sequence, as well as substitutions of single bases.

D. CYTOPLASMIC INHERITANCE: In recent years, evidence has begun to accumulate which suggests the presence of hereditary systems outside the chromosomes of the cell nucleus. The pattern which appears to be emerging is that cytoplasmic structures (such as chloroplasts and mitochondria) contain separate genes which are regulated by the nuclear genes.

Investigations of "killer" strains of *Paramecium* and the recent discovery of small amounts of DNA in chloroplasts may be described in support of cytoplasm inheritance.

ACTION OF GENES IN PEAS

AIM: To show that differences which seem to be just gross may well be microscopic and chemical as well. Thus the biochemical and enzymatic differences of the peas are hereditary and of importance in many ways in the matter of breeding.

MATERIALS NEEDED: Thirty smooth peas; 30 wrinkled peas; two small bottles; medicine droppers; mortar and pestle; razor blade; petri dishes; balance; compound microscope; slides; cover glasses; paper towels; distilled water; nutrient agar; iodine solution for starch tests; blendor; funnel; filter paper; cheesecloth; and 1 gram glucose-l-phosphate.

PROCEDURE:

(A) Gross Differences:

1. Weigh the 30 smooth peas and then the 30 wrinkled peas. Record their weight. Now place the peas into 2 bottles, 1 bottle labeled for the smooth peas and 1 bottle labeled for the wrinkled peas. Fill the bottles almost full of water and put aside until the next day.

2. The following day, remove the peas from the water and dry them as well as possible with paper towels. Now weigh them and record the weight

after soaking 24 hours. Determine the weight increase in both sets of seeds and calculate the percentage of increases in weight.

3. What can you now say about the appearance of both kinds of peas and the ability of each to soak up water?

(B) Microscopic Differences:

1. Take a microscope slide and mark one-half of the slide with an S for smooth and the other with a W for wrinkled. Place a drop of water on either side of the slide.

2. Cut through a soaked smooth pea with a razor blade and scrape a little of the cut surface into the drop of water on the half of slide marked S. Do the same with the soaked wrinkled pea. Mix thoroughly with the water.

3. Use the low-power of the microscope to examine the scrapings. Then use the high-power. Describe and draw what you see. There should be a definite difference in the starch grains with the smooth seed having smooth undivided starch grains as contrasted with the divided dimpled starch grains of the wrinkled peas.

(C) Chemical Differences

1. Make a 2% agar solution which should contain ½% of glucose-l-phosphate. One grain is sufficient for 20 petri dishes. Boil the agar and the glucose until the foam looks coarse. You do not have to sterilize the mixture but pour into petri dishes to form a layer about 5 mm deep. Allow to gel. Cover the dishes and refrigerate until needed.

2. Weigh out 12 grams of dry smooth peas and 12 grams of dry wrinkled peas.

3. Either grind each group to a fine powder with a mortar and pestle or else place into a blendor for a few seconds to get a fine powder.

4. Now place the powder into a bottle, add 12 ml of distilled water and mix thoroughly. Filter through several layers of cheesecloth. Label each bottle either smooth or wrinkled filtrate.

5. Take the petri dishes, divide the plate into two halves by drawing a line with a wax pencil on the outside of the bottom dish. Mark one side S and the other W.

6. Taking a medicine dropper, place 4 separate drops of the extract from the smooth seeds on the side marked S and then place 4 separate drops of the extract from the wrinkled seeds on the side marked W. If you have an incubator, place the dishes in for 30 minutes. Otherwise allow them to stand for 30 minutes on the table. This is so that the extract can act on the glucose.

7. At the end of the time, add a drop of iodine to 2 of the drops on each side of the dish. You should find that a blue ring has formed at the drops of the smooth seed extract and a very dark blue ring at the wrinkled seed extract. Now wipe off the extract and iodine with filter paper.

8. Allow the dishes to stand for another 30 minutes. Test the remaining

2 drops on each side of the dish with the iodine solution. There should be very little or no starch found in the smooth peas while there still are traces of starch in the extract of the wrinkled peas.

OBSERVATIONS AND CONCLUSIONS:

Since we are assuming that both the smooth and the wrinkled peas have been dried under similar conditions, then the reason for the differences in the gross features, the type of starch grains found, and the enzymes controlling the synthesis of carbohydrates must come from the genetic differences of the peas.

1. After studying this, what would you say is probably the best observation to distinguish the primary action of the gene?
2. Is the information obtained from this exercise of any importance?

UNIT IX

EVOLUTION AND DIVERSITY

I. EVIDENCES FOR EVOLUTION

Study of past and present organisms indicates that many new species have arisen during the long period of the earth's history. Many species exhibit similarities which suggest common ancestry.

Evidences supporting the concept of organic evolution as a reason for the similarities and diversities among species can be observed in the study of: the fossil record, comparative anatomy, comparative embryology, comparative biochemistry, and the geographic distribution of organisms.

A. *The Fossil Record*: The preserved remains of many organisms have been recovered from rock layers which were formed during the remote history of the earth. Fossil characteristics suggest a pattern of relationships which had undergone a process of consecutive change throughout long periods of time.

To use the fossil record as evidence for evolution it must be assumed that, in undisturbed layers of the earth's crust, the oldest layers are the lowest ones, and each succeeding layer is "younger" than the layer on which it is superimposed. Acceptance of this assumption is the basis for studying serial records of fossils.

B. *Comparative Anatomy*: A comparative study of certain organisms shows striking similarities in anatomical features. In many instances skeletal comparisons suggest development from a common ancestral form, along evolutionary lines reflecting the particular environment in which the related organs function. Such homologous structures include:

- the forelegs of frogs and horses
- the flippers of whales and the arms of man
- the wings of bats and birds.

C. *Comparative Embryology*: Comparison of embryonic structures among groups of organisms indicates similarities which strongly suggest common ancestry.

- similarities of vertebrate embryos in early development.
- Resemblance of some mollusc larvae to annelids.

D. *Comparative Biochemistry*: Biochemical similarities among organisms also support the idea of common ancestry.

- the closer the relationship among organisms the greater the similarity of nucleic acids.
- many related organisms have the same enzymes and hormones.

E. *Geographic Distribution*: Geographic isolation of organisms over the earth increases the likelihood of genetic differentiation.

- Darwin's finches
- Abundance of marsupials in Australia
- similarities and differences of Old World and New World monkeys.

II. THEORIES OF EVOLUTION

Men have studied the variety of life for centuries. As a result, various theories have been developed to explain the cause of the diversity of organisms, as well as the reasons for their adaptations (structures related to functions). Several of these theories have had great impact on evolutionary thought.

A. *Lamarck's Theory of Use and Disuse*: Lamarck believed that new organs arise according to the needs of an organism and that the size of the organ is due to the degree to which it is used or not used. He thought that useful characteristics acquired in one generation may be transmitted to the next.

B. *Darwin's Theory of Natural Selection*: Darwin published his theory after many years of observation and careful deduction. Darwin's theory includes these contributory ideas:

- More offspring are produced than can possibly survive.
- Since the number of adults in populations tends to remain constant from generation to generation, a struggle for survival is suggested.
- The individuals that survive are the ones best fitted to exist in their environment, due to differences (variations) in their characteristics.
- Individuals that survive tend to reproduce their kind, transmitting their favorable variations to their offspring.
- As time and generations continue, the favorable changes (adaptations) are perpetuated in the species.

C. *Modern Theory*: Modern evolutionary theory retains the basic essentials of the Darwinian theory, and has added additional information resulting from investigations which followed Darwin's work. The years since the publication of Darwin's theory have seen great advances in the understanding of the genetic basis of evolutionary change.

1. *Production of Variations*: The genetic basis for variations within a species is apparently provided by gene mutations, chromosome mutations, and the sorting out and recombination of genes associated with sexual life cycles.

2. *Population Genetics*: In recent years geneticists have applied the principles of Mendelian genetics to the study of the behavior of genes in populations of sexually reproducing organisms.

a. Population: A population includes all members of a species inhabiting a given location.

b. Gene Pool: The gene pool of a population consists of the sum total of all the heritable genes for all the traits in a given population.

c. The Hardy-Weinberg Principle: Hardy and Weinberg extended the principles of Mendelian genetics to populations of sexually reproducing individuals and made the following observations:

• The gene pool of a population tends to remain stable (i.e., gene frequencies remain constant) if certain conditions are met, including

- large populations
- random mating
- no migration
- no mutations.

• After one generation of random mating, the genotypes in a population will tend to distribute according to the relationship, $p^2 + 2pq + q^2$. In this relationship, p = the frequency of the dominant allele (A), and q = the frequency of the recessive allele (a). Thus, p^2 = the frequency of homozygous dominant (AA) individuals, $2pq$ = the frequency of heterozygous (Aa) individuals, and q_2 = the frequency of homozygous recessive (aa) individuals in the population.

3. *Transmission of Variations*: The constancy of a population suggested by the Hardy-Weinberg principle rarely occurs since some or all of the conditions for gene pool stability may not be met. The unstable gene pool that results tends to produce change in the population. The change that results is dependent on the pattern of transmission of variations that arises in the population as acted upon by the environment of the individuals in the populations.

a. Factors Influencing Transmission: The principal forces influencing the pattern of change in a population are the environment and isolation.

(1) *The Environment*: The distribution of alleles in a population tends to reflect the adaptive value of the traits controlled by the alleles. Traits which offer high survival value to individuals possessing them tend to increase in frequency. Traits, and the alleles associated with them, that

have low adaptive value tend to have a low frequency of occurrence.

If environmental conditions change, selective forces then act on the gene pool of a population in a different manner, and gene frequencies (hence, traits) may change markedly. Traits that formerly had low survival value may, in a changed environment, have greater survival value and increase accordingly. Conversely, high frequency traits may diminish considerably (even disappear) in a changed environment.

(2) *Isolation*: Some members of a population may become separated from the original, continuous population. If a population becomes subdivided, transmission of variations is not likely to follow the same pattern in the subdivisions.

- Mutations are not likely to be the same.
- Random recombination is not likely to occur in the same manner.

Geographic isolation, reproductive isolation, and isolation in time prevent the intermingling of traits within a species and tend to maintain differences.

b. Patterns of Change: Species tend to change in time as random events (mutations and recombination) occur against a background of environmental change and isolation.

(1) *Change within a Species*: Throughout time most species have changed as environmental conditions have varied and selective forces have increased or diminished the survival value of various traits. Examples:

- occurrence of antibiotic-resistant strains of bacteria
- occurrence of DDT-resistant house flies

(2) *Development of New Species*: According to present thinking, new species evolve as a result of the interaction of the following factors:

- variations
- changes in existing environments
- migrations to new environments
- adaptation to changed or new environments
- natural selection
- isolation from other species with which they might interbreed if not isolated.

When members of one species are geographically isolated, selection tends to act upon separate patterns of randomness to produce divergent trends which may ultimately result in changed individuals. Such changed organisms may become incapable of interbreeding and producing fertile offspring. When such reproductive isolation occurs, two species are said to exist.

The pattern of species formation is not thought to be entirely random. Since new species stand a better chance of survival if they are not in direct competition with existing species, they tend to evolve so as to occupy those environmental conditions which are least utilized in the area where they exist. Such a pattern of evolution, where species tend to develop to fill the available environmental niches, is said to be a pattern of adaptive radiation. The evolutionary pattern of Darwin's finches is an excellent example of adaptive radiation.

III. A MORE MODERN CLASSIFICATION SYSTEM

The great diversity of organisms present on the earth today is the result of billions of years of evolution. Biologists, in attempting to "sort out" and comprehend relationships among this vast number of different types, have found it essential to devise systems by which organisms may be classified. A useful classification system would reflect evolutionary similarities and differences, but since so much of the evolution story is not clearly understood, there is some disagreement as to the best classification system.

In one modern system, organisms are grouped into three kingdoms, each of which is subdivided into phyla (singular, phylum) which are categories indicating major differences in organization and function.

The three kingdoms are Protist, Plant, and Animal.

Protists:

- Protozoa
- Algae
- Fungi
- Slime molds
- Bacteria

Plants:

- Bryophytes
- Tracheophytes

Animals:

- Sponges
- Coelenterates
- Flatworms
- Roundworms
- Segmented worms
- Molluscs
- Arthropods
- Echinoderms
- Chordates
    - Fish
    - Amphibians
    - Reptiles
    - Birds
    - Mammals

IV. ORIGIN AND EARLY EVOLUTION OF LIFE

Modern evolutionary theory also includes an explanation for the origin and early evolution of life on the primitive earth.

A. *The Heterotroph Hypothesis*: The heterotroph hypothesis is an explanation for the origin of life, which holds that the first life was not able to synthesize its own organic nutrient materials from inorganic compounds. Like all scientific explanations of incompletely understood phenomena, the heterotroph hypothesis is based on logical extensions of certain basic assumptions.

1. *Assumptions*:

a. Raw Materials: It is assumed that the primitive earth was an exceptionally hot body consisting of inorganic compounds in solid, liquid, and gaseous states, and a rich environmental supply of energy.

(1) Inorganic Compounds: Water, condensing and falling as rain, carried dissolved gases and minerals to form seas in depressions on the cooling earth's crust. In addition to dissolved minerals, the primitive oceans contained concentrations of dissolved ammonia, methane, and hydrogen.

(2) Energy sources: In addition to heat, energy in the form of lightning discharges from the atmosphere, solar radiation (including x-rays and ultraviolet rays) and "hard" radiation from crystal rocks offered an energy-rich environment.

b. Synthesis: Available energy permitted the formation of chemical bonds among dissolved particles in the "hot thin soup" of the seas. Such random synthesis led to the formation of organic molecules such as simple sugars and amino acids. These organic molecules interacted to form more complex organic molecules.

c. Some of the large, complex molecules formed aggregates (clusters). These aggregates probably incorporated molecules from the seas as "food" (heterotrophic nutrition).

d. Reproduction: In time, as aggregates became increasingly complex and highly organized, the ability to reproduce evolved. The aggregates are considered to have been alive when they evolved reproductive capacity.

B. *Heterotroph to Autotroph*: It is thought that heterotrophs evolved a pattern of respiration similar to the anaerobic process of fermentation. Extended periods of fermentative activity would have added quantities of carbon dioxide to the environment. Some heterotrophs evolved a means of using the carbon dioxide to synthesize organic compounds. These were the pioneer autotrophs.

C. *Anaerobe to Aerobe*: Autotrophic activity tended to add oxygen to the environment. Some autotrophs and heterotrophs evolved mechanisms for using oxygen for deriving energy (aerobic respiration). Modern organisms may be heterotrophic aerobes or anaerobes, or autotrophic aerobes or anaerobes. The evolution from the first life to the great variety and complexity of life that exists today required billions of years of slow change.

BIOGENESIS

AIM: To demonstrate that life comes only from life.

Going back to the time of the ancients it was believed that there was spontaneous generation. Aristotle was the first to expound this theory. With time it was seen that larger plants and animals, in fact all living organisms, arose directly from other living things.

However, after van Leeuwenhoek discovered bacteria in 1676, the controversy again arose concerning the origin of these minute living things. This argument was not settled among scientists until Pasteur, through a series of experiments in the 1860's, showed that bacteria, too, must have parents.

Before Pasteur did his classical experiment there were two other scientists who attempted to prove the same thing. The first one was Redi and the second one was Spallanzani.

In this exercise we shall attempt to prove the same thing by doing all three experiments.

However, to simplify the laboratory techniques, each of the three will be done individually with each one listed as a separate exercise as follows:

REDI'S EXPERIMENT

MATERIALS NEEDED: Three jars; 3 pieces of fresh meat; a piece of cheesecloth or closely woven mesh; a piece of plastic or parchment.

PROCEDURE:

Place a piece of meat in each of the three jars. Cover one with the mesh or cheesecloth, one with the plastic or parchment and leave the third uncovered. Place the jars outdoors or in a warm place. Within a few days, you will find eggs of blowflies on the meat in the open jar and on the cheesecloth of the second jar, with no trace of flies on the jar with the plastic. This shows that if there was no way for the odor of decaying meat to escape, the flies were not able to reach it and leave eggs there. Although the flies were not able to reach the meat itself in the jar covered with the cheesecloth, the flies still attempted to get to it and laid the eggs on the mesh. As for the third jar, the flies had no difficulty in getting at the meat itself.

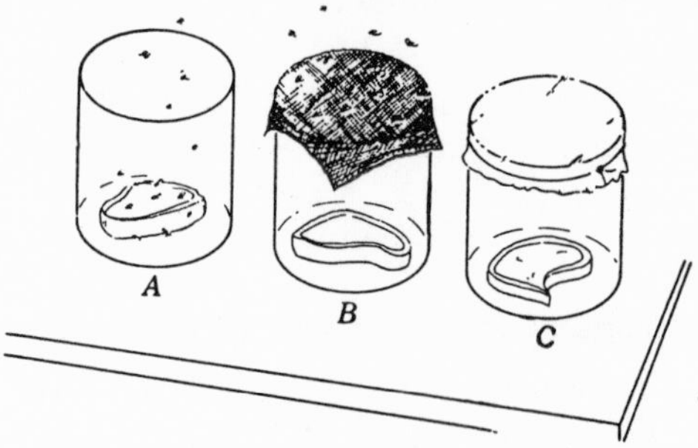

Diagram 36. Redi's Experiment.

SPALLANZANI'S EXPERIMENT

MATERIALS NEEDED: Three test tubes; prepared beef broth; water bath; cotton plugs; Bunsen burner; test tube rack; wing top for burner.

PROCEDURE:

1. Take the 3 test tubes and make a constriction about 3/4 of an inch in length in the middle as illustrated (Diagram 37).

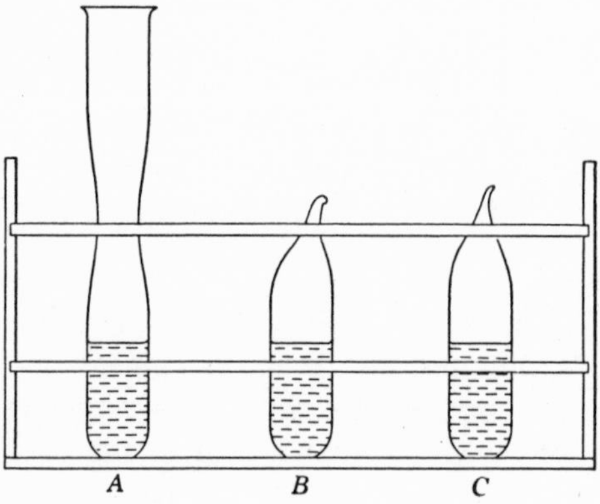

Diagram 37. Spallanzani's Experiment.

2. Take cool beef broth and fill the lower part of each tube with broth. Now arrange and label the tubes as shown in the illustration.
3. Tube A should be boiled in a water bath and then left unsealed.

4. Tube B should be in boiling water until the broth boils and then quickly plugged with cotton. When the test tube is cool enough to handle, hold it at a 45 degree angle over the Bunsen burner and heat the constricted area. Pull out the top and seal the tube.

5. Seal tube C just as you did tube B but do not boil the broth.

6. Place the tubes into a warm place. In a few days you will find that tubes A and C contain a broth that is turbid while tube B is completely clear.

PASTEUR'S EXPERIMENT

MATERIALS NEEDED: Three small Erlenmeyer flasks; prepared beef froth; water bath; S-shaped glass tube; straight glass tube; 2 one-hole stoppers; sterilizer; graduate cylinders.

PROCEDURE:

(1) Put 10 ml of the broth in each of the 3 sterilized flasks and mark them A, B, and C.

(2) In flask A, place a rubber stopper with a sterilized straight tube. Place in boiling water bath.

(3) In flask B, place a rubber stopper with a sterilized S-tube. Place in boiling water bath.

(4) Flask C should be left unstoppered and unsterilized.

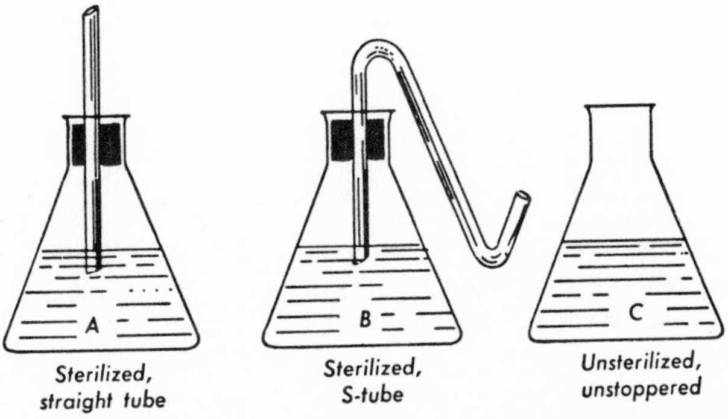

Diagram 38. Pasteur's Experiment.

(5) After 20 minutes in the water bath, place all three flasks in a warm place.

(6) Within a few days you will notice that the broth in flask C will become cloudy, it will have an odor, and a surface film.

(7) At the same time flask A should show a slight degree of contamination while flask B should show no contamination.

OBSERVATIONS AND CONCLUSIONS:

1. Why would you say that Spallanzani's experiments were more valid than those of Redi?
2. In Pasteur's experiment, why would you say that flask B remained uncontaminated?
3. Why should flask A in Pasteur's experiment become contaminated since the flask and the tube were both sterilized before the broth was put in?
4. How could you prove in Pasteur's experiment that everything was sterile, including the broth at the beginning of the experiment and that there were bacteria in flask A and C at the end?

VARIATIONS

AIM: To demonstrate that traits will vary within a species.

You have learned that biological traits will show variations. In this exercise you will measure and examine several different items to verify this.

MATERIALS NEEDED: For each study: 20 leaves off the same tree; ½ lb. of fresh green peas in pods; ½ lb. of lima or any other dried beans; rulers.

PROCEDURE:

(1) Each student should carefully measure the size of the beans in his ½ lb. and record the results.

(2) Each student should open the pea pods and examine each pod for difference in size, shape, number and placement of the peas. Record the results.

(3) Each student should carefully examine each of the leaves for size, shape, color, vein formation, and outer edge indentations.

OBSERVATIONS AND CONCLUSIONS:

1. It will be demonstrated that no two things are exactly alike even though they might be of the same species, or growing on the same tree. How can this variation be accounted for?
2. Why is it normal for a member of a species to be different from other members?
3. What are the two major causes of variations among members of a species?
4. How could one determine whether a variation is acquired or inherited?

BIOCHEMICAL EVIDENCES OF EVOLUTION

AIM: To show that the pattern of amino acids differs as we go up the evolutionary scale.

Since it is simplest to obtain certain marine invertebrates, you will use samples of these animals to show variations in amino acids that are to be found in several phyla of lower animals.

The best method of tracing the various types of amino acids found in these animals is by means of chromatography and comparison with ordinary amino acids. You have done Thin-Layer Chromatography of Amino Acids and will use the technique and materials of that exercise in the work done in this exercise.

MATERIALS NEEDED: All the materials listed in the amino acid chromatography; pieces of fresh muscle from scallops, oysters, clams, octopuses, lobsters, shrimp, snails, star-fish, any fairly large coelenterate such as a sea anemone; 80% ethyl alcohol; mortar and pestle; graduates; beakers; Erlenmeyer flasks; funnels; centrifuge; filter paper.

PROCEDURE:

The students are to work in teams, each student being assigned to a particular piece of muscle and one student being assigned to do the TLC of the amino acids. Thus the number of students in each of the teams will depend upon the number of specimens being examined.

1. Each student is to place his particular tissue into a mortar containing 10 ml of alcohol. Grind the tissue thoroughly; decant the liquid into a beaker. Add another 20 ml of the alcohol and grind again. Then combine the decanted liquids and ground tissues into a flask. Label. Stopper and leave in the refrigerator over night.
2. The following morning centrifuge the contents of the beaker. Filter the liquids into a beaker and evaporate the liquid by means of fans until about 10 ml remain.
3. Now make a thin-layer chromatogram of the colorless liquid.
4. Compare your chromatogram with the chromatogram of the amino acids to determine whether your pattern varies. Compare your chromatogram with the ones prepared by your teammates, each of whom was using a different animal specimen.

OBSERVATIONS AND CONCLUSIONS:

1. Make up a table showing which amino acids are found in which animals.
2. If you were to correlate the results of the chromatogram in terms of evolution, what would you say in regards to amino acid patterns?
3. Are there any definite pattern evidences as the animal is found higher up on the "tree of life"?
4. Why would you expect to find the results you have?

CLASSIFICATION

AIM: To become familiar with the methods of classifying plants and animals.

Organisms are generally classified into two great kingdoms, Animal and Plant, while many biologists group the single-celled organisms into a third kingdom, known as the Protista.

Each kingdom is then subdivided into phyla which are large groups consisting of those organisms that have common characteristics. The phyla are further subdivided into classes, the classes into orders, the orders into families, the families into genera, and the genera into species.

In this exercise we shall not classify the organisms past the outstanding characteristics which are the basis for the classification.

MATERIALS NEEDED: Three species of plants and three species of animals for a group of 4 students. Try to vary the specimens as you hand them out to the students so that there will be many different specimens on display for study and the students can then pool their results.

PROCEDURE:

Each student should examine the 6 specimens he has for his personal study and if time permits, the student should study the other specimens scattered throughout the room.

For the plant studied, the student should answer the following questions for each of the plants:

1. Does the plant contain chlorophyll?
2. Does the plant have roots, stems, and leaves?
3. Does the plant have flowers?
4. Is the plant single-celled, treelike, or water-growing?
5. To what phyla does the plant belong and why?
6. What characteristics of the phyla does this plant have?
7. Are there any other reasons why this plant should be classified in the manner you have done?

For the animal study the student should answer the following questions for each of the animals:

1. Does the animal have several body layers?
2. Does the body surface have numerous pores?
3. Is the body plan radial or bisymmetrical?
4. Is the body wormlike?
5. Does the animal have an external covering?
6. Is the animal a vertebrate or invertebrate? Why?
7. What phylum does this animal belong in?
8. Do you know the common name of this animal?

9. What characteristics of the animal are similar to animals in general?

10. What characteristics of the animal are different from those of animals in general?

OBSERVATIONS AND CONCLUSIONS:

1. Would you say that within a phylum, all classes should have the same characteristics of the phylum?
2. Would you say that the classes within a phylum should be more similar to one another than to classes belonging to another phylum?
3. Do you think that animals can be classified on the basis of their habits or their geographical distribution?

UNIT X

PLANTS AND ANIMALS IN THEIR ENVIRONMENT

I. ECOLOGY

Ecology is the study of the interactions of plants and animals, and their interrelationships with the physical environment. No organism exists as an entity, separate and distinct from its environment, since a dynamic equilibrium exists in the interactions of organisms with themselves and with the environment.

A. *Physical Factors in the Environment*: Various physical and chemical factors affect the ability of an organism to live and reproduce in any location.

These nonliving, or abiotic, factors include:

—light
—temperature
—moisture
—substratum (soil or rock)
—available inorganic substances, such as minerals
—oxygen supply.

B. *Biotic Factors in the Environment*: The interactions of organisms (biotic factors) influence their development and survival and that of the species.

The term "organism" implies a prime characteristic of the living world —ORGANIZATION—which is evident from the atom to the biosphere.

The concept of organization is apparent in:

—atoms —inorganic molecules —large organic molecules
—nucleic acids —chromosomes —viruses —cells
—tissues —organs —organ systems —organisms
—populations —communities —ecosystems
—world biomes —biosphere

II. BIOTIC ORGANIZATION

A. *Population*: All members of a species inhabiting a given location constitute a group of organisms that are more or less alike. They are capable of interbreeding and producing fertile offspring under natural conditions.

B. *Community*: A unit composed of all plant and animal populations interacting in a given environment is known as a community.

C. *Ecosystem*: The living community and nonliving environment function together as an ecological system or an ecosystem.

1. *Interactions*: An ecosystem involves interaction among organisms, as well as a flow of energy and a cycling of materials. *Examples*: A balanced aquarium, a self-contained space ship.

a. Specific relationships: Various species of organisms that live in the same area, and are thus part of the same ecosystem, affect each other in a variety of ways. This may be illustrated by the interactions involving nutrition.

Organisms which cannot synthesize their own food are said to be dependent, meaning dependent upon other organisms for food. This dependency is shown by:

—saprophytes, the plants which live on dead organic matter. These include the decomposers upon which all organisms depend, directly or indirectly. *Examples*: bacteria of decay, mushrooms.

—herbivores, the plant-eating animals. *Examples*: cows, horses.

—carnivores, the flesh-eating animals. These include:

—predators, animals which kill other living animals and feed on them. *Examples*: lions, hawks.

—scavengers, animals which eat organisms that they have not killed. *Examples*: vultures, hyenas.

Some animals may be both scavengers and predators as, for example, the snapping turtle.

—omnivores, animals that eat both plants and animals. *Example*: man.

Different organisms may live together in a close association which may or may not be beneficial to them. This living together in intimate association is known as symbiosis.

Types of symbiosis include:

—commensalism—one organism is benefited and the other is not affected. *Examples*: Remora and shark, barnacle and whale.

—mutualism—both organisms benefit from the association. *Examples*: alga and fungus in a lichen, nitrogen-fixing bacteria and legumes.

—parasitism—the parasite benefits at the expense of the host. *Examples*: athlete's foot fungus and man; lamprey eel and trout.

Competition occurs when organisms living in the same environment utilize the same limited resource, such as food and space, and other environmental factors such as water, light, oxygen, and minerals.

The more similar the requirements of the organisms involved, the more intense is the competition.

If two different species compete for the same food or reproductive sites, one or the other, due to differences in reproductive rates, will be successful in eliminating the other. This establishes one species per niche in a balanced (climax) community. All members of that species use the same kinds of food and occupy similar reproductive sites. The term habitat refers to where organisms live; the term niche refers to their role in the biotic community.

b. General Relationships

(1) *Energy Flow*: Activity is characteristic of life and requires the expenditure of energy.

The complex pathways of energy through the living components of the ecosystem are represented in the concept of the food chain and web.

Energy from the sun enters living systems through green plants. A food chain involves the transfer of food energy from the source in green plants through a series of organisms with repeated stages of eating and being eaten.

In a natural community the flow of energy (and materials) is really much more complicated than is suggested by any one food chain, since practically all species of organisms may be consumed by more than one other species. Therefore, there are so many branchings and cross-branchings among the food chains of any community that the whole situation is better termed a food web.

Balance is achieved in a food web by interactions among:

—producers—All of the food for a community is derived from organic compounds synthesized in green plants. Green plants are, therefore, considered to be producers.

—consumers—Organisms that feed directly upon green plants (herbivores) are known as primary (or first-level) consumers. Secondary (or second-level) consumers (carnivores) prey upon primary consumers and gain their food and energy from them.

—decomposers—Living organisms concentrate and organize materials of the environment into complex living substances. During their lifetime they give off wastes. Wastes and dead organisms are eventually broken down to simpler compounds by decomposers. These compounds are thus returned to the environment where they can be used by other living organisms.

In going from the producer through the various members of a food web, the total amount of energy is decreased, since each member of the web utilizes some of this energy in consuming source material and for its own metabolism, and gives off additional energy in the form of heat.

Since there is a loss of energy at each feeding level, the producer organisms contain the greatest amount of energy. Energy is steadily expended and a system cannot continue functioning without the constant input of energy from the outside.

Consumers are usually larger than the organisms they eat. As a consequence, many organisms at a lower feeding level are required to furnish sufficient materials and energy to supply a single organism at a higher feeding level.

(2) *Material Cycles*: Material is cycled between organisms and their environment, passing from inorganic sources to organic forms and back again to inorganic forms. This is accomplished chiefly through the action of decomposers, scavengers, and saprophytes.

The Nitrogen Cycle: Nitrogen is found in the protoplasm of living things during protein synthesis. The nitrogen of protein is released and converted to nitrates by the sequential activities of certain bacteria (the decay and nitrifying bacteria). Nitrogen-fixing bacteria also produce nitrates from gaseous nitrogen. Thus, nitrogen in the form of nitrates is made available to plants for protein synthesis. Animals eat plants and synthesize animal protein from plant protein. Plants and animals produce nitrogenous wastes during their lifetime and become sources of energy for the decomposers (bacteria of decay) when they die. Denitrifying bacteria contribute to the cycle by returning molecular nitrogen to the atmosphere.

2. *Maintenance*: An ecosystem is self-sustaining if there is:

—a constant source of energy

—a living system capable of incorporating this energy into organic compounds

—a cycling of materials between organisms and their environment.

3. *Changes*: A self-perpetuating community in which populations exist in balance with each other and with the environment is said to be stable. Such a stable, or climax, community—the living part of an ecosystem—persists until major climatic, geologic, or biotic changes alter or destroy it, thus producing non-climax conditions.

Thereafter, a community goes through a series of successive stages until climax conditions are reached.

Each community modifies the environment, making it more unfavorable for itself, and, apparently, more favorable for the following community.

This replacement of one community by another until a climax stage is reached is known as ecological succession.

Successive communities are composed of populations able to exist under the prevailing conditions and are identified by their dominant species —the one which exerts control over the other species present.

Dominant plant species also dominate in the sense that they are the most abundant food present. Animal populations are limited to those which can use these plants (or each other) for food.

a. Pioneer Organisms: Succession may be said to begin with pioneer organisms, since these are the first plants to populate a given location. Lichens, for example, are pioneer organisms on bare rock. Since the pioneer organisms change the environment, habitats for other organisms are developed.

b. Climax Community: The climax community is not immediately replaced; long periods of time are required for ecological succession after a major disturbance. In New York State, for instance, climax communities vary. At higher elevations hemlock-beech-maple associations are often the climax flora. At lower elevations, oak-hickory associations are frequently the climax flora. Some disturbances result in changes so drastic that the original climax community never returns.

D. *World Biomes*: A biome represents a major ecological grouping of organisms.

1. *Terrestrial*: The major plant and animal associations on land are determined by the major climate zones of the world, modified by local land and water conditions. Climatic factors including temperature, precipitation, and solar radiation, plus other factors, such as the nature of the earth's crust, determine the habitats for living organisms.

Land biomes are characterized and named for the climax vegetation in a region. Since green plants are the producers of food, the major plant associations of the earth determine the animals that can live in any locality.

The major land biomes are:

—tundra
—taiga (coniferous forest)
—temperate deciduous forest
—tropical forest
—grassland
—desert

Increase in latitude or in altitude can bring about similar climatic conditions which results in similar plant associations at higher latitudes and higher altitudes.

2. *Aquatic*: The criteria used to classify terrestrial biomes are not applicable to large bodies of water, such as oceans, where the distribution of plants does not exert a controlling influence.

The temperature variation is not great and moisture is seldom limiting in aquatic areas, which represent the largest and most stable ecosystem on earth.

More than 70 percent of the earth's surface is covered by water, and most life on the planet exists in conditions where water is the principal external medium.

Such physical factors as temperature, the quantity of available oxygen and carbon dioxide, and suspended or dissolved materials are the major ones affecting aquatic organisms. Aquatic organisms which carry on aerobic respiration utilize atmospheric oxygen which has penetrated and dissolved into the water medium. The quantity of oxygen present in any body of water limits the quantity of aerobic life that may live there.

The greatest amount of food production in the world occurs in the oceans, along the edges of land masses (coastal waters), wherever water is fairly clear.

a. *Marine*: The world ocean is a huge, continuous body of water that:

—absorbs and holds large quantities of solar heat and regulates the earth's temperature

—contains a relatively constant supply of nutrient materials and dissolved salts.

—serves as a habitat for most organisms.

The salts of the oceans, dissolved from the land masses, not only furnish nutrients to organisms, but are related to their water balance. In general, salt concentrations in marine environments are very similar to salt concentrations in living cells.

The quantity of light available for photosynthesis in the oceans is determined by the intensity of radiation entering the water, quantity of reflection, and transparency of the water.

Water absorbs much light energy, so that most photosynthesis takes place near the surface of the oceans, the deeper regions being too dark.

b. *Fresh Water*: As in marine environments, available respiratory gases, temperature, transparency, and depth of the water, as well as salt concentration, affect life in fresh water.

Due to the lower concentration of salts in fresh water, a diffusion gradient exists between cells of organisms and their water environment.

Water balance in fresh water animals is maintained by regular removal of excess water by means of specialized structures (a homeostatic mechanism).

E. *Biosphere*: The biosphere is that portion of the earth in which ecosystems operate, and includes the biologically inhabited air, soil, and water.

The term "biosphere" also includes the relationships among living things for the transfer of materials and energy.

III. BIOSPHERE AND MAN

A. *Past and Present*: Man exerts a unique and powerful influence on the physical and living world. He has been enormously successful in modifying his environment.

His failures, however, can be attributed to the lack of realization that he not only influences other individuals, other species, and the nonliving world, but is influenced by them.

1. *Negative Aspects*: The delicate balance of nature has been upset as a result of:

Man's Numbers: The population total of humans has risen at a rapid rate partly because of the removal of natural checks on populations. This increase in human population has far outstripped the food-producing capacity of many ecosystems of the world.

Man's Activities: Increasing urbanization has claimed increasing amounts of food-producing land.

Man has permitted overgrazing, overcropping, and overburning, resulting in the loss of topsoil cover in areas so treated. Removal of soil (and forest) cover has led to disastrous floods.

Technological activities have polluted two components of the environment on which life depends: air and water. Escalating demands for water (agriculture, sewage disposal, industrial use, hydroelectric power, and recreation) are being made on a supply which is being increasingly polluted. Burning and other industrial processes are the major causes of air pollution.

Major water pollutants include: raw sewage, chemicals, fertilizers, silt, and salt water.

Man has brought about the extinction or near extinction of many animal species, not only through direct killing, but also through human alteration of habitat. Extinct species include: passenger pigeon, great auk, dodo, Carolina paroquet, and heath hen. Nearly extinct species include: California condor, grizzly bear, American crocodile, Blue whale.

The decline of the Kaibab deer population from 1907 to 1925 illustrates the effects of the removal of normal checks (predators) on species populations.

Species of plants also face extinction because of destruction or uncontrolled disease. Examples include: American elm, American chestnut.

The use of pesticides (in an attempt to control certain pests) has compounded the pest problem, contaminated the atmosphere, and water supply, and disrupted food webs. The decline of the bald eagle may possibly be attributed to disrupted food webs.

In the case of insects, mutant species resistant to insecticides have increased faster than new insecticides can be developed. Since insects are man's chief competitors for available food, an increase in their numbers represents formidable competition. It has been established that forage crops can absorb insecticides from the soil, thus entering food chains, and webs.

The use of antibiotics in an attempt to control certain pathogens has contributed to the rise of resistant mutant species of pathogens, and has thus compounded the disease problem.

Man has imported organisms, such as the starling and rabbit, which have unexpectedly led to the disruption of existing ecosystems. Global travel increases the dissemination to new environments of pathogens, and insect and plant pests. Examples of imported pests include: the corn borer, Japanese beetle, and Hessian fly.

2. *Positive Aspects*: In some cases man has successfully used ecological knowledge:

—Methods of controlling the human reproductive rate are being developed.

—The sea is being extensively investigated as a source of food. Since many soil nutrients lost to crop plants are carried out to sea, return by way of marine food may assist in perpetuating cycles.

—Reforestation, soil cover plantings, and dam construction serve as flood and soil erosion controls in many areas. Soil reclamation measures have returned otherwise lost soil to use.

—Attempts are being made in large cities, like New York, to control air and water pollution by laws and by the development of new techniques of sanitation.

—Man has attempted to preserve some plants and animals which are or were on the verge of extinction. Game preserves, national parks, hunting laws, and fisheries, represent some efforts by man to preserve species. Animals which were near extinction and which are now increasing in numbers as a result of protection are the bison, egret, and trumpeter swan. The future of some species, such as the California condor, whooping crane, and the redwoods, is still in doubt.

—Biological control of insect pests based on extensive studies of the life histories of these pests is now being encouraged. This is less likely to affect those species of insects which are helpful to man than is the use of insecticides. The undesirable effects of pesticides, in general, may possibly be avoided by the use of carefully considered biological controls. *Example*: the use of cobalt-60 radiation on male screw worm flies.

—Some imported animals, such as the Italian honeybee and the Chinese pheasant, have proved extremely valuable in their transplanted environments.

—Control of human infectious diseases has led to an increase in the life span of man.

—Space exploration includes ecological considerations.

B. *The Future*: The lack of understanding on the part of man has led to the ecological problems of the past and present. The future rests on man's acknowledgment that the principles of ecology which apply to plants and other animals apply equally to him.

ECOLOGY AROUND YOUR SCHOOL

AIM: To learn the principles of ecology by observing the plant and animal life found around your school on its grounds.

All organisms exist in a state of equilibrium with other organisms and their environment. Usually most of these interactions can be seen in a small area. Your school has some open space surrounding it. If this area is examined, you would be surprised at the number of ecological interrelationships and principles you will discover.

MATERIALS NEEDED: An outline map of the school grounds; funnels of one size; beakers of one size; 100 ml graduate cylinder; ring stands and rings; microscope slides; cover glasses; methylene blue; absorbent cotton.

PROCEDURE:

1. First walk around the area and record the density and location of plants on the outline map. Use the symbols: T for trees, S for shrubs, P for non-woody plants, and G for grass.
2. You will also record any rocks that might be on the grounds by using R.
3. If you see any birds, add the letter B and record in which areas they were seen.
4. You might find some insects and worms on the grounds. Record the locations with an I for insects and a W for worms.
5. Now that you have recorded everything that was obvious to see, turn over the rocks (do this as a class) and record the kind of animals that you found there. Also, what were these animals doing?
6. As a class bring back to the classroom some samples of soil from various parts of the grounds and from around clover plant roots. Check the soils for their water-retaining properties by placing each of the samples into a funnel over a piece of absorbent cotton. This cotton has been placed into the opening of the funnel to prevent the soil from being washed through. The funnel should be placed in the ring on the ring stand and the water allowed to go through into the beaker. Record the time it takes for 100 ml of water to filter through, from the time the first drop is poured into the funnel until the last is filtered through.
7. Make a slide of a very dilute solution of soil dug up around clover plants. Stain with methylene blue. Under low and high power of the microscope look for any organisms.

OBSERVATIONS AND CONCLUSIONS:

1. Have the plants and trees used as landscape been placed in their position for any specific reasons?
2. Where the plants have been growing wild, what have you noticed about their growth?

3. Where have you found the birds?
4. What do the worms and insects do for the soil?
5. Were there any invertebrates under rocks? If so, why?
6. What organisms have you observed in the soil water? Are they of any particular value in this ecological interrelationship?
7. What is the relationship between the composition of the soil and the types of plant and animal life it supports?
8. The organisms found in the soil around the clover roots are part of what cycle?
9. Where in this exercise do you have part of the nitrogen cycle in operation?
10. Where in this exercise, do you have the principle of soil water and the various types of soils?

POND WATER LIFE

AIM: What kind of organisms can be found in a drop of pond water?

Few people realize that a pond is a functioning ecosystem. It contains producers, consumers, predators, scavengers and decomposers.

However, these organisms are spread throughout the pond water and in order to concentrate the organisms for you to see, it is necessary that the sample of pond water be concentrated. This is done by means of the centrifuge which is a high-speed rotary machine for separating solids from liquids or liquids of different specific gravities from each other. Since the organisms are the solid part of the water, they will tend to fall to the bottom of the centrifuge tubes in a small amount of water.

MATERIALS NEEDED: Pond water concentrated through centrifugation; medicine dropper; microscope slide; silicone gum; cover glass; microscope; identification charts and books; methyl cellulose.

PROCEDURE:

1. Use the silicone gum and slide to make a living cell. Place a drop of the pond water into the cell and cover with the cover glass. Focus under the low power of the microscope and have the field as dark as possible so that the organisms can be seen.
2. Observe carefully and draw as many of the organisms as you are able. If possible identify by means of the various charts and books.
3. Put aside the slide with the living cell to be observed again in about 24 hours. At that time look at the organisms again and see whether you still find the same organisms. Some of them may have been used as food by the others.

4. After observing the organisms the first time, place a drop of the pond water on another slide, place the cover glass on and draw through a minute quantity of the methyl cellulose to slow down the organisms by means of a paper towel. Observe under high power. Record your observations.

5. If you have time make another slide from the top of the water and note whether the organisms are similar to those on the previous slide.

OBSERVATIONS AND CONCLUSION:

1. Is there any difference between the sample from the surface and the sample from the bottom?
2. Are there both plants and animals in the sample?
3. What did you find in the sample that you had in the living cell for 24 hours?
4. Would you say that there is a food chain in the pond?
5. Do you think that this is typical of other ecosystems?

MINIATURE OXYGEN-CARBON-DIOXIDE CYCLE

AIM: To demonstrate the oxygen–carbon-dioxide cycle in a closed ecosystem. Plants take in carbon dioxide and give off oxygen in photosynthesis, and animals take in oxygen and give off carbon dioxide in respiration.

Usually this cycle is seen in a large balanced aquarium which contains small fish, some fresh-water snails, crayfish, tadpoles, salamanders, water beetles, etc., and water plants such as Elodea, Vallisneria, and Nitella. However, in this exercise we shall demonstrate that the cycle operates regardless of the size of the ecosystem. Test tubes have also been used for this purpose but here is a closed ecosystem that can be made on an ordinary glass microscope slide.

MATERIALS NEEDED: Concentrated pond water; medicine dropper; microscope slide; silicone gum; cover glass; plant such as Nitella; microscopic worms such as Nais or Dero, and a crustacean such as Daphnia; microscope.

PROCEDURE:

1. Make a fairly large-size silicone gum living cell.
2. Into the cell place the growing tip and a small segment of Nitella which is an aquarium plant.
3. Add several drops of concentrated pond water so that there are protozoa in the cell.
4. In addition, place some microscopic worms such as Nais or Dero and a small crustacean, such as Daphnia, into the cell.

5. Apply the cover glass and look at the material under a microscope or use a microprojector.

6. Keep the microaquarium in moderate light and examine daily. You will see that all the organisms stay alive and well.

7. As controls, prepare two other cells, one without the plant and one with the plant, in plain distilled water. Observe for several days.

OBSERVATIONS AND CONCLUSIONS:

1. Why do you find that the slide containing a balanced life remains in operation?

2. What would you expect to find when there is no plant present?

3. What would you expect to find where there is no animal present?

4. You might find that this closed system will only work properly for a period of time and then decomposition will take place. Can you think of the reason?

SOIL BACTERIA AND THE NITROGEN CYCLE

AIM: To study the part played by soil bacteria in the nitrogen cycle.

There is approximately 80% nitrogen in air; but since very few organisms are capable of using the nitrogen directly from the atmosphere, it must be changed into a form that can be used by living organisms.

Nitrogen-fixing bacteria absorb free nitrogen from the air and convert it into soluble nitrates usable by plants for making proteins. The nitrogen-fixing bacteria live in nodule-like swellings in the roots of leguminous plants. At the same time there are putrefactor bacteria which take the dead organic matter found in soil and convert it to ammonia. This ammonia (NH_3) is converted to nitrite (NO_2) and finally to nitrate (NO_3) by various bacterial forms present in the soil.

In this very simple exercise you will see that bacteria found in soil will cause nitrogen compounds to change to ammonia.

MATERIALS NEEDED: Two test tubes; 4% peptone solution; soil; incubator; slides; medicine dropper; Nessler's reagent; Hydrion paper.

PROCEDURE:

Nessler's reagent is an indicator for ammonia while peptone solution is a solution rich in polypeptides.

1. Fill both test tubes with the peptone solution.

2. Into tube A place some soil dug up from around plant roots while leaving tube B without any soil.

3. Incubate both tubes for 48 hours.

4. Place a drop of Nessler's reagent on each of the slides.

5. Add a drop of the liquid from tube A on one slide and a drop of the liquid from tube B on the other slide.

6. By means of a toothpick mix the reagent and the liquid together on each of the slides and wait a few minutes.

7. Are there any color changes?

8. Test the pH of both tubes with the help of Hydrion paper.

OBSERVATIONS:

1. Which tube contained the liquid which turned yellow? Why did this occur?

2. From which part of the protein molecules did the NH_3 originate?

3. When you checked the pH of both tubes, what did you learn?

4. Continuing on with the pH, why don't all organisms give off nitrogen waste in the form of ammonia?

5. In this exercise you used an artificial protein source; what would be the natural protein source in the soil?

SALINITY AND LIVING ORGANISMS

AIM: To determine the effects of salinity on living organisms.

To aquatic organisms, salinity is of great importance. It has been found that the percentage of salts in natural waters varies greatly. If the water has no outlet such as an inland lake, there is more salt to be found in the water than if the water has an outlet such as a river.

As a result, the organisms found in each of these waters will differ since the tolerance to salinity varies among the various species of organisms.

Since sodium chloride is the salt found most in natural waters, it will be used in this exercise.

To determine the effects of the salt concentrations on various organisms, we shall use both plants and animals that are usually aquatic.

MATERIALS NEEDED: Living specimens of aquatic organisms such as Elodea, Nitella, Spirogyra, small snail, Daphnia, Hydra, Planaria; medicine dropper; microscope slides; cover glasses; compound microscope; paper towels; watch; distilled water or tap water; sodium chloride solutions of 1, 3, 5, and 10% strengths; pond water; depression slides.

PROCEDURE:

1. Place a drop of pond water on a slide and cover with a cover glass. Examine under low and high power to determine the normal reactions of the protists.

2. With the slide still on the microscope, add a drop of 1% salt solution to one edge of the cover slip. Draw the salt solution under the cover slip by placing a small piece of toweling at the opposite edge of the cover glass. Record the time. Observe under low power. Is there any reaction in the organisms you saw before?

3. Draw off the water and replace with tap water to determine the time it takes for the organism to recover.

4. Follow steps 2 and 3 for each of the salt concentrations.

5. If the organisms died, record the time it took and the lethal concentration.

6. Do this for all your specimens. If the organism is too large to be placed flat on a microscope slide, place the organism in the depression slide and cover with the cover glass. Then to remove the solutions and replace with those of greater concentrations, use a medicine dropper. Test each organism for recovery time in ordinary tap water.

OBSERVATIONS AND CONCLUSIONS:

1. Make a table in which you list the organism, the concentration of the saline solutions, the response to the concentrations and the recovery time.

2. If the organism did not recover, what kind of aquatic habitat do you think it normally inhabited?

3. Did all the organisms react in a similar manner?

4. Which of the organisms was most tolerant of the various changes in the salt concentrations and which of the organisms was least tolerant?

5. Which of the organisms you have worked with would be most widely distributed in the waters of the world?

6. Which of the organisms must remain in one location because it is not tolerant of changes in saline concentrations?

ECOLOGICAL SUCCESSION

AIM: To study succession in a microcosm and to study evidences of succession in several communities.

Usually we think of succession as the transition from one type of fauna to another within a given area; the major phases being herbaceous vegetation, shrubs, and finally forest, with minor transitions in each phase. Another definition of the same word means the changes in the population of an area, as a result of the response of competing organisms to the environment.

In this exercise we are dealing with various populations living in balance with each other and with the environment. However, as each new organism gains supremacy the environment changes to favor that species. Thus populations in environments such as ponds tend to change from hour to hour and from day to day. Conditions that can cause these changes are too much rain, too little rain, too much acidity, too much sun, and no sun.

We shall attempt to study some of these environmental changes in true pond water and simulated pond community.

MATERIALS NEEDED: Five quart jars without covers and one jar with a cover; cultures of pond water; microscope; Hydrion paper; medicine droppers; centigrade thermometer; slides; cover glasses; dried plant material from pond; sterile pond water.

PROCEDURE:

1. For the simulated pond water, boil pond water and place 3/4 full into a clear jar with a cover. Add a handful of dried grass, leaves, or other plant material which you obtained from the margin of the pond. Cover the jar.
2. From the same pond from which you obtained the water for Step 1, fill 5 quart jars 3/4 full of water containing some of the plant materials. Try to have one jar contain only algae, another only duckweed, a third the submerged vegetation and dead material, the fourth water and material from the edge of the pond, and the fifth, if possible, material from the center of the pond.
3. You now have 6 different environments. Place the cultures in the area of the laboratory where all the jars will receive the same amount of light at all times.
4. Check the temperature of the water in all 6 jars, every day.
5. Check the pH of the water each day to see whether it varies. Use the Hydrion paper.
6. Every third day, make microscopic slides of the material in the jars. Record any changes in organisms found in the jars.
7. Record any changes that you can see with the naked eye as far as the cultures are involved.
8. Make a chart of all the living organisms that you have observed for a period of three weeks. List when you have observed them, how long they lived, under what conditions they thrived, and under what conditions they died.

OBSERVATIONS AND CONCLUSIONS:

1. Why did you cover the jar and boil the pond water in jar #1?
2. Why is temperature important for succession?
3. How is the pH of the culture related to succession? If there were changes, what caused them?
4. What causes the changes in the water?
5. What organisms assumed dominance after the water became turbid?

Where did these organisms come from?

6. What evidence did you see of population changes within the community?

7. Did a food chain or food web exist? Explain.

8. Did your different communities finally reach a stable or climax condition? How would you know?

9. Was there any real difference in the succession between the covered sterile jar and the other 5 communities?

10. On the basis of your results, set up the course of succession in a fresh-water culture and discuss the factors that govern it.

CONCLUSION

These are just a very few experiments that can be performed in a high school biology laboratory. However, the author feels that the ones selected for use in this book were best equipped to further the understanding of modern biological principles by a student in high school.

The author hopes that the teacher who uses the book will share this opinion.

INDEX